THE HOUR
BETWEEN DOG
AND WOLF

THE HOUR
BETWEEN DOG
AND WOLF

Risk-Taking, Gut Feelings
and the Biology of Boom and Bust

JOHN COATES

RANDOM HOUSE CANADA

PUBLISHED BY RANDOM HOUSE CANADA

www.randomhouse.ca

Random House Canada and colophon are registered trademarks.

Library and Archives Canada Cataloguing in Publication

Coates, John M.

The hour between dog and wolf : risk taking, gut feelings
and the biology of boom and bust / John Coates.

Includes bibliographical references.

Also issued in electronic format.

ISBN 978-0-307-35967-4

1. Decision making—Physiological aspects. 2. Risk-taking (Psychology)—
Physiological aspects. 3. Neuroeconomics. 4. Cognitive neuroscience. I. Title.

QP360.5.C62 2012   612.8'233   C2011-908176-8

Typeset in Minion by G&M Designs Limited, Raunds, Northamptonshire

Figures on pp. 46, 69, 103, 126 and 199 by CLIPAREA.com – Custom Media. Figure 1 (p. 32)
from Wikimedia Commons. Figure 2 (p. 37) from Nick Hobgood, Wikimedia Commons.
Figure 4 (p. 61) from Kate from UK, Wikimedia Commons.

Cover design by Tal Goretsky

Printed and bound in the United States of America

10 9 8 7 6 5 4 3 2 1

For Ian, Eamon, Iris and Sarah

CONTENTS

[The hour] between dog and wolf, that is, dusk, when the two can't be distinguished from each other, suggests a lot of other things besides the time of day ... The hour in which ... every being becomes his own shadow, and thus something other than himself. The hour of metamorphoses, when people half hope, half fear that a dog will become a wolf. The hour that comes down to us from at least as far back as the early Middle Ages, when country people believed that transformation might happen at any moment.

JEAN GENET, *PRISONER OF LOVE* (1986, TRANS. BARBARA BRAY)

PART I

Mind and Body in the
Financial Markets

INTRODUCTION

When you take risks, you are reminded in the most insistent manner that you have a body. For risk by its very nature threatens to hurt you. A driver speeding along a winding road, a surfer riding a monster wave as it crests over a coral reef, a mountain climber continuing his ascent despite an approaching blizzard, a soldier sprinting across no-man's land – each of these people faces a high chance of injury, even death. And that very possibility sharpens the mind and calls forth an overwhelming biological reaction known as the 'fight-or-flight' response. In fact, so sensitive is your body to the taking of risk that you can be caught up in this visceral turmoil when death poses no immediate threat. Anyone who plays a sport or watches from the stands knows that even when it is 'just a game', risk engages our entire being. Winston Churchill, a hardened campaigner from the most deadly wars, recognised this power of non-lethal risk to grip us, body and mind. When writing of his early years, he tells of a regimental polo match played in southern India that went to a tie-break in the final chukka: 'Rarely have I seen such strained faces on both sides,' he recalls. 'You would not have thought it was a game at all, but a matter of life and death. Far graver crises cause less keen emotion.'

Similar strong emotions and biological reactions can be triggered by another form of non-lethal risk – financial risk-taking. With the exception of the occasional broker suicide (and these may be more myth than reality), professional traders, asset managers and individuals investing from home rarely face death in their dealings. But the bets they place can threaten their job, house, marriage, reputation and social class. In this way money holds a special significance in our lives. It acts as a powerful token distilling many of the threats and opportunities we faced over eons of evolutionary time, so making and losing it can activate an ancient and powerful physiological response.

In one important respect, financial risk carries even graver consequences than brief physical risk. A change in income or social rank tends to linger, so when we take risks in the financial markets we carry with us for months, even years after our bets have settled, an inner biological storm. We are not built to handle such long-term disturbances to our biochemistry. Our defence reactions were designed to switch on in an emergency and then switch off after a matter of minutes or hours, a few days at the most. But an above-average win or loss in the markets, or an ongoing series of wins or losses, can change us, Jekyll-and-Hyde-like, beyond all recognition. On a winning streak we can become euphoric, and our appetite for risk expands so much that we turn manic, foolhardy and puffed up with self-importance. On a losing streak we struggle with fear, reliving the bad moments over and over, so that stress hormones linger in our brains, promoting a pathological risk-aversion, even depression, and circulate in our blood, contributing to recurrent viral infections, high blood pressure, abdominal fat build-up and gastric ulcers. Financial risk-taking is as much a biological activity, with as many medical consequences, as facing down a grizzly bear.

This statement about biology and the financial markets may sound strange to ears accustomed to the teachings of economics. Economists tend to view the assessment of financial risk as a purely intellectual affair – requiring the calculation of asset returns, probabilities, and the optimal allocation of capital – carried on for the most part rationally. But to this bloodless account of decision-making I want to add some guts. For recent advances in neuroscience and physiology have shown that when we take risk, including financial risk, we do a lot more than just think about it. We prepare for it physically. Our bodies, expecting action, switch on an emergency network of physiological circuitry, and the resulting surge in electrical and chemical activity feeds back on the brain, affecting the way it thinks. In this way body and brain twine as a single entity, united in the face of challenge. Normally this fusion of body and brain provides us with the fast reactions and gut feelings we need for successful risk-taking. But under some circumstances the chemical surges can overwhelm us; and when this happens to traders and investors they come to suffer an irrational exuberance or pessimism that can destabilise the financial markets and wreak havoc on the wider economy.

To give you a mere inkling of how this physiology works, I am going to take you onto the trading floor of a Wall Street investment bank. Here we will observe a high-stakes world where young bankers can step up or down a full social class in the space of a single year, one year buying a beach house in the Hamptons, the next pulling their kids out of private school. So consider if you will the following scenario, in which an unanticipated and important piece of news impacts an unsuspecting trading floor.

INCOMING!

It has been said of war that it consists of long stretches of boredom punctuated by brief periods of terror, and much the same can be said of trading. There are long stretches of time when little more than a trickle of business flows in through the sales desks, perhaps just enough to keep the restless traders occupied and to pay the bills. With no news of any importance coming across the wire, the market slows, the inertia feeding on itself until price movement grinds to a halt. Then, people on a trading floor disappear into their private lives: salespeople chat aimlessly with clients who have become friends, traders use the lull to pay bills, plan their next ski trip, or talk to headhunters, curious to know their value on the open market. Two traders, Logan, who trades mortgage-backed bonds, and Scott, who works down the aisle on the arbitrage desk, toss a tennis ball back and forth, taking care not to hit any salespeople.

This afternoon the Federal Reserve is holding a meeting of its Board of Governors, and normally these events are accompanied by market turbulence. It is at these meetings that the Fed decides whether to raise or lower interest rates, and should it do so it announces its decision at 2.15 p.m. Even though the economy has been growing at a healthy clip and the stock market has been unseasonably, even irrationally, strong, the Fed has dropped few hints of an increase. So today it is widely expected to leave rates unchanged, and by late morning most people across the trading floor haven't a worry on their minds, and think of little else but whether to order sushi or pasta for lunch.

But just before noon there comes the merest breath of change, rippling the surface of prices. Most people on the floor do not consciously notice it, but the slight tremor registers none the less. Maybe their breathing quickens, maybe muscles tense just a bit, maybe arterial blood pressure increases ever so slightly. And the sound of the floor shifts, from the quiet buzz of desultory conversation to a mildly excited chatter. A trading floor acts as a large parabolic reflector, and through the bodies of its thousand-odd traders and salespeople it gathers information from faraway places and registers early signals from events that have yet to happen. The head of the trading floor looks up from his papers and steps out of his office, surveying the floor like a hunting dog sniffing the air. An experienced manager can sense a change in the market, tell how the floor is doing, just from the sight and sound of it.

Logan stops in mid-throw and looks over his shoulder at the screens. Scott has already wheeled his chair back to his desk. Their monitors display thousands of prices and flowing news feeds, blinking and disappearing. To outsiders the vast matrix of numbers seems chaotic, overwhelming, and finding the significant bit of information in the mess of prices and irrelevant news items seems as impossible as picking out a single star in the Milky Way. But a good trader can do just that. Call it a hunch, call it gut feeling, call it tradecraft, but this morning Scott and Logan have sensed a kaleidoscopic shift in price patterns well before they can say why.

One of the brain regions responsible for this early-warning system is the locus ceruleus (pronounced ser-u-leus), so called because its cells are cerulean, or deep blue. Situated in the brain stem, the most primitive part of the brain, sitting atop the spine, the locus ceruleus responds to novelty and promotes a state of arousal. When a correlation between events breaks down or a new pattern emerges, when something is just not right, this primitive part of the brain registers the change long before conscious awareness. By doing so it places the brain on high alert, galvanising us into a state of heightened vigilance, and lowering our sensory thresholds so that we hear the faintest sound, notice the slightest movement. Athletes experiencing this effect have said that when caught up in the flow of a game they can pick out every voice in the stadium, see every blade of grass. And today when the stable correlations between asset prices broke down the locus ceruleus tripped an alarm, causing Scott and Logan to orient to the disturbing information.

Moments after Scott and Logan have pre-consciously registered the change, they learn that one or two people on the Street have heard, or suspected, that the Fed will raise interest rates this afternoon. Such a decision announced to an unprepared financial community would send a tidal wave of volatility through the markets. As the news and its implications sink in, Wall Street, only a short while ago looking forward to calling it an early day, roils with activity. At hastily organised meetings traders consider the possible Fed moves – will it leave rates unchanged? Raise them a quarter of a percentage point? Half a per cent? What will bonds do under each scenario? What will stocks do? Having formed their views, traders then jostle to set their positions, some selling bonds in anticipation of a rate hike, which pushes the market down almost 2 per cent, others buying them at the new lower levels, convinced the market is oversold.

Markets feed on information, and the Fed announcement will be a feast. It will bring volatility to the market, and volatility to a trader means a chance to make money. So this afternoon most traders exude excitement, and many of them will make their entire week's profit in the next few hours. Around the world bankers stay up to hear the news, and trading floors now buzz with a ludic atmosphere more commonly found at a fair or sporting event. Logan warms to the challenge and with a rebel yell dives into the seething market, selling $200 million mortgage bonds, anticipating an exciting ride down.

By 2.10, trading on the screen dwindles. The floor goes quiet. Across the world traders have placed their bets, and now wait. Scott and Logan have readied their positions and feel intellectually prepared. But the challenge they face is more than an intellectual puzzle. It is also a physical task, and to perform it successfully they require a lot more than cognitive skills. They also need fast reactions, and stamina enough to support their efforts for the hours ahead when volatility spikes. What their bodies need, therefore, is fuel, lots of it, in the form of glucose, and they need oxygen to burn this fuel, and they need an increased flow of blood to deliver this fuel and oxygen to gas-guzzling cells throughout the body, and they need an expanded exhaust pipe, in the form of dilated bronchial tubes and throat, to vent the carbon dioxide waste once the fuel is burned.

Consequently Scott and Logan's bodies, largely unbeknownst to them, have also prepared for the event. Their metabolism speeds up, ready to break down existing energy stores in liver, muscle and fat cells should the situation demand it. Breathing accelerates, drawing in more oxygen, and their heart rates speed up. Cells of the immune system take up position, like firefighters, at vulnerable points of their bodies, such as the skin, and stand ready to deal with injury and infection. And their nervous system, extending from the brain down into the abdomen, has begun redistributing blood throughout their bodies, constricting blood flow to the gut, giving them the butterflies, and to the reproductive organs – since this is no time for sex – and shunting it to major muscle groups in the arms and thighs as well as to the lungs, heart and brain.

As the sheer potential for profit looms in their imaginations, Scott and Logan feel an unmistakable surge of energy as steroid hormones begin to turbo-charge the big engines of their bodies. These hormones take time to kick in, but once synthesised by their respective glands and injected into the bloodstream, they begin to change almost every detail of Scott and Logan's body and brain – their metabolism, growth rate, lean-muscle mass, mood, cognitive performance, even the memories they recall. Steroids are powerful, dangerous chemicals, and for that reason their use is tightly regulated by law, by the medical profession, by the International Olympic Committee, and by the hypothalamus, the brain's 'drug enforcement agency'; for if steroid production is not turned off quickly it can transform us, body and mind.

From the moment the rumour first spread, and over the past couple of hours, Scott and Logan's testosterone levels have been steadily climbing. This steroid hormone, naturally produced by the testes, primes them for the challenge ahead, just as it does athletes preparing to compete and animals steeling for a fight. Rising levels of testosterone increase Scott and Logan's haemoglobin, and consequently their blood's capacity to carry oxygen; the testosterone also increases their state of confidence and, crucially, their appetite for risk. For Scott and Logan, this is a moment of transformation, what the French since the Middle Ages have called 'the hour between dog and wolf'.

Another hormone, adrenalin, produced by the core of the adrenal glands located on top of the kidneys, surges into their blood. Adrenalin quickens physical reactions and speeds up the body's metabolism, tapping into glucose deposits, mostly in the liver, and flushing them into the blood so that Scott and Logan have back-up fuel supplies to support them in whatever trouble their testosterone gets them into. A third hormone, the steroid cortisol, commonly known as the stress hormone, trickles out of the rim of the adrenal glands and travels to the brain, where it stimulates the release of dopamine, a chemical operating along neural circuits known as the pleasure pathways. Normally stress is a nasty experience, but not at low levels. At low levels it thrills. A non-threatening stressor or challenge, like a sporting match, a fast drive or an exciting market, releases cortisol, and in combination with dopamine, one of the most addictive drugs known to the human brain, it delivers a narcotic hit, a rush, a flow that convinces traders there is no other job in the world.

Now, at 2.14, Scott and Logan lean into their screens, gaze steady, pupils dilated; heart rates drop to a slow idle; their breathing rhyth-

mic and deep; muscles coiled; body and brain fused for the impending action. An expectant hush descends on global markets.

THE INSIDE STORY

In this book I tell the story of Scott and Logan, of Martin and Gwen,
and of a trading floor of supporting characters, as they are caught in
the floodtide of a bull and then a bear market. The story will consist of
two threads: a description of overt trading behaviour – how professional traders make and lose money, the euphoria and stress that
accompany their changing fortunes, the calculations behind bonus
payments – and a description of the physiology behind the behaviour.
The threads will, however, lace together, forming a single story. Splicing
the two will enable us to see how brain and body act as one during
important moments in a risk-taker's life. We will explore pre-conscious
circuits of the brain and their intimate links with the body in order to
understand how people can react to market events so fast that their
conscious brain cannot keep up, and how they draw on signals from
the body, the fabled gut feelings, to optimise their risk-taking.

Despite the traders' frequent successes, the story follows the narrative arc of tragedy, with its grim and unstoppable logic of overconfidence and downfall, what the ancient Greeks called *hubris* and
nemesis. For human biology obeys seasons of its own, and as traders
make and lose money they are led almost irresistibly into recurrent
cycles of euphoria, excessive risk-taking and crash. This dangerous
pattern repeats itself in the financial markets every few years. Alan
Greenspan, former chairman of the US Federal Reserve, puzzled over
this periodic folly, and wrote of 'innate human responses that result
in swings between euphoria and fear that repeat themselves generation after generation'. Much the same pattern occurs in sport, politics
and war, where larger-than-life characters, believing themselves
exempt from the laws of nature and morality, overreach their abilities.
Extraordinary success seems inevitably to breed excess.

Why is this? Recent research in physiology and neuroscience can, I
believe, help us explain this ancient, delusional and tragic behaviour.

Human biology can today help us understand overconfidence and irrational exuberance, and it can contribute to a more scientific understanding of financial market instability.

A simpler reason for bringing biology into the story is that it is, quite simply, fascinating. A story of human behaviour spiked with biology can lead to particularly vivid moments of recognition. The term 'recognition' is commonly used to describe the point in a story when all of a sudden we understand what is going on, and by that very process understand ourselves. It was Aristotle who coined this term, and since his day recognition moments have been largely the preserve of philosophy and literature. But today they are increasingly provided, for me at any rate, by human biology. For when we understand what is going on inside our bodies, and why, we are met with repeated Aha! moments. These range from the fun: 'Oh, so that's why I get butter-flies in my stomach when excited!' or 'So that's why I get goosebumps when scared!' (The erector pili muscles in your skin try to raise your fur, to make you look bigger, just as a cat does when threatened. Most of your fur no longer exists, so you get goosebumps instead, but where it does you have a 'hair-raising' experience) – to the deadly serious: 'So that's why stress is so tormenting, why it contributes to gastric ulcers, hypertension, even heart disease and stroke!'

Today human biology, perhaps more than any other subject, throws a light into the dark corners of our lives. So by mixing biology into the story I can more accurately describe what it feels like to take large financial risks; and I can do so moreover in a way that provides recog-nition moments for people who have never set foot on a trading floor. In fact, the physiology I describe is not confined to traders at all. It is the universal biology of risk-taking. As such it has been experienced by anyone who plays a sport, runs for political office, or fights in a war. But I focus on financial risk-taking, and do so for good reason: first, because finance is a world I know, having spent twelve years on Wall Street; second, and more importantly, because finance is the nerve centre of the world economy. If athletes succumb to overconfi-dence, they lose a match, but if traders get carried away on a flood of hormones, global markets founder. The financial system, as we have recently discovered to our dismay, balances precariously on the mental health of these risk-takers.

I begin by looking at the physiology that produces our risk-taking, filling in the background story for what follows. I then show, through a story set on a trading floor, how this physiology can mix with lax risk-management systems and a bonus system that rewards excessive gambling to produce a volatile and explosive bank. We watch as nature and nurture conspire to produce an awful train wreck, leaving behind mangled careers, damaged bodies and a devastated financial system. We then linger in the wreckage and observe the resulting fatigue and chronic stress, two medical conditions that blight the workplace. Finally we look at some tentative yet hopeful research into the physiology of toughness, in other words training regimes designed by sports scientists and stress physiologists to immunise our bodies against an overactive stress response. Such training could help calm the unstable physiology of risk-takers.

ONE

The Biology of a Market Bubble

THE FEELING OF A BUBBLE

My interest in the biological side of the financial markets dates back
to the 1990s. I was then working on Wall Street, trading derivatives for
Goldman Sachs, then Merrill Lynch, and finally running a desk for
Deutsche Bank. This was a fascinating time to trade the markets,
because New York, and indeed America as a whole, was caught up in
the dot.com bubble. And what a bubble that was. The markets had
not seen anything quite like it since the great bull market of the 1920s.
In 1991 the Nasdaq (the electronic stock exchange where many new-
tech ventures are listed) traded below 600, and had meandered around
that level for a few years. It then began a gradual yet persistent bull
run, reaching a level of 2,000 in 1998. The Nasdaq's rise was checked
for a year or so by the Asian Financial Crisis, which reined it back
about 500 points, but then the market recovered and took to the skies.
In little more than a year and a half the Nasdaq shot up from 1,500 to
a peak just over 5,000, for a total return in excess of 300 per cent.

The rally was almost unprecedented in its speed and magnitude. It
was completely unprecedented in the paucity of hard financial data
supporting the dot.com and high-tech ventures powering the bull
run. In fact, so large was the gap between stock prices and the underlying fundamentals that many legendary investors, betting unsuccess-
fully against the trend, retired from Wall Street in disgust. Julian
Robertson, for instance, founder of the hedge fund Tiger Capital,
threw in the towel, saying in effect that the market may have gone

crazy but he had not. Robertson and others were right that the market was due for a dreadful day of reckoning, but they also fully understood a point made by the great economist John Maynard Keynes back in the 1930s: that the markets could remain irrational longer than they, the investors, could remain solvent. So Robertson retired from the field, his reputation and capital largely intact. Then, early in 2000, the Nasdaq collapsed, giving back over 3,000 points in little more than a year, eventually bottoming out at the 1,000 level where it had begun a few short years before. Volatility of this magnitude normally makes a few people rich, but I know of no one who made money calling the top of this market's explosive trajectory.

Besides the scale of the run-up and subsequent crash, another feature of the Bubble was noteworthy, and reminiscent of the 1920s, at least the 1920s I knew from novels, black-and-white movies and grainy documentaries – that was how its energy and excitement overflowed the stock exchange, permeated the culture and intoxicated people. For the fact is, while they last, bubbles are fun; and the widespread silliness attending them is often remembered with a certain amount of humour and fondness. I imagine anyone who lived through the bull market of the Roaring Twenties retained an abiding nostalgia for that heroic and madcap time, when futuristic technology, blithe spirits and easy wealth seemed to herald a new era of boundless possibility. Of course, life in its aftermath must have been even more formative, and those born and raised during the Great Depression are said to carry, even into old age, what the historian Caroline Bird calls an 'invisible scar', a pathological distrust of banks and stock markets, and a morbid fear of unemployment.

My recollections of the 1990s are of a decade every bit as hopeful and every bit as screwball as the 1920s. During the nineties we were entertained by middle-aged CEOs in black turtleneck sweaters trying to 'think outside the box'; by kids in their twenties wearing toques and yellow sunglasses, backed by apparently limitless amounts of capital, throwing lavish parties in midtown lofts and talking wacky internet schemes few of us could understand – and even fewer questioned. To do so meant you 'just didn't get it', one of the worst insults of the time, indicating that you were a dinosaur incapable of lateral thought. One thing I definitely didn't get was how the internet was supposed to overcome the constraints of time and space. Sure, ordering online was easy, but then delivery took place in the real world of rising oil prices and road congestion. The internet company that made the most heroic attempt to defy this brute fact was Kozmo.com, a New York-based start-up that promised free delivery within Manhattan and about a dozen other cities within an hour. The people who paid the price for this act of folly, besides the investors, were the scores of bicycle messengers breathlessly running red lights to meet a deadline. You would see groups of these haggard youngsters outside coffee bars (with appropriate names like Jet Fuel) catching their breath. Not surprisingly the company went bankrupt, leaving behind a question asked about this and countless similar ventures: what on earth were the investors thinking?

Perhaps the right question should have been, were they thinking at all? Were investors engaged in a rational assessment of information, as many economists might – and did – argue? If not, then were they perhaps engaged in a different form of reasoning, something closer to a game theoretic calculation: 'I know this thing is a bubble,' they may have schemed, 'but I'll buy on the way up and then sell before everyone else.' Yet when talking to people who were investing their savings in newly listed internet shares I found little evidence for either of these thought processes. Most investors I spoke to had difficulty employing anything like linear and disciplined reasoning, the excitement and boundless potential of the markets apparently being enough to validate their harebrained ideas. It was almost impossible to engage them in a reasoned discussion: history was irrelevant, statistics counted for little, and when pressed they shot off starbursts of trendy concepts like 'convergence', the exact meaning of which I never discerned, although I think it had something to do with everything in the world becoming the same – TVs turning into phones, cars into offices, Greek bonds yielding the same as German, and so on.

If investors who had bought into this runaway market displayed little of the thought processes described by either rational choice or game theory, they also displayed little of the behaviour implied by a more common and clichéd account – the fear and greed account of investor folly. According to this piece of folk wisdom a bull market, as it picks up steam, churns out extraordinary profits, and these cause the better judgement of investors to become warped by the distemper of greed. The implication is that investors know full well that the market is a bubble, yet greed, rather than cunning, causes them to linger before selling.

Greed certainly can and does cause investors to run with their profits too long. By itself, though, the account misses something important about bubbles like the dot.com era and perhaps the Roaring Twenties – that investors naïvely and fervently believe they are buying into the future. Cynicism and cunning are not on display. Furthermore, as a bull market starts to validate investors' beliefs, the profits they make translate into a lot more than mere greed: they bring on powerful feelings of euphoria and omnipotence. It is at this point that traders and investors feel the bonds of terrestrial life slip from their shoulders and they begin to flex their muscles like a newborn superhero. Assessment of risk is replaced by judgements of certainty – they just *know* what is going to happen: extreme sports seem like child's play, sex becomes a competitive activity. They even walk differently: more erect, more purposeful, their very bearing carrying a hint of danger: 'Don't mess with me,' their bodies seem to say. 'I can handle anything.' Tom Wolfe nailed this delusional behaviour when he described the stars of Wall Street as 'Masters of the Universe'.

It was this behaviour more than anything else that struck me during the dot.com era. For the undeniable fact was, people were changing. The change showed itself not only among the untrained public but also, perhaps even more, among professional traders all along Wall Street. Normally a sober and prudent lot, these traders were becoming by small steps euphoric and delusional. Their minds were frequently troubled by racing thoughts, and their personal habits were changing: they were making do with less sleep – clubbing till 4 a.m. – and seemed to be horny all the time, more than usual at any rate, judging by their lewd comments and the increased amount of porn on their computer screens. More troubling still, they were becoming overconfident in their risk-taking, placing bets of ever-increasing size and with ever worsening risk–reward trade-offs. I was later to learn that the behaviour I was witnessing showed all the symptoms of a clinical condition known as mania (but now I am getting ahead of the story).

These symptoms are not unique to Wall Street: other worlds also manifest them, politics for example. One particularly insightful account of political mania has been provided by David Owen, now Lord Owen. Owen, a former British Foreign Secretary and one of the founders of the UK Social Democratic Party, has spent most of his life at the very top of British politics. But he is by training a neurologist, and has lately taken to writing about a personality disorder he has observed among political and business leaders, a disorder he calls the Hubris Syndrome. This syndrome is characterised by recklessness, an inattention to detail, overwhelming self-confidence and contempt for others; all of which, he observes, 'can result in disastrous leadership and cause damage on a large scale'. The syndrome, he continues, 'is a disorder of the possession of power, particularly power which has been associated with overwhelming success, held for a period of years and with minimal constraint on the leader'. The symptoms Owen describes sound strikingly similar to those I observed on Wall Street, and his account further suggests an important point – that the manic behaviour displayed by many traders when on a winning streak comes from more than their newly acquired wealth. It comes equally, perhaps more, from a feeling of consummate power.

During the dot.com years I was in a good position to observe this manic behaviour among traders. On the one hand I was immune to the siren call of both Silicon Valley and Silicon Alley. I never had a deep understanding of high tech, so I did not invest in it, and could watch the comedy with a sceptical eye. On the other, I understood the traders' feelings because I had in previous years been completely caught up in one or two bull markets myself, ones you probably did not hear about unless you read the financial pages, as they were isolated in either the bond or the currency market. And during these periods I too enjoyed above-average profits, felt euphoria and a sense of omnipotence, and became the picture of cockiness. Frankly, I cringe when I think about it.

So during the dot.com bubble I knew what the traders were going through. And the point I want to make is this: the overconfidence and hubris that traders experience during a bubble or a winning streak just does not feel as if it is driven by a rational assessment of opportunities, nor by greed – it feels as if it is driven by a chemical.

When traders enjoy an extended winning streak they experience a high that is powerfully narcotic. This feeling, as overwhelming as passionate desire or wall-banging anger, is very difficult to control. Any trader knows the feeling, and we all fear its consequences. Under its influence we tend to feel invincible, and to put on such stupid trades, in such large size, that we end up losing more money on them than we made on the winning streak that kindled this feeling of omnipotence in the first place. It has to be understood that traders on a roll are traders under the influence of a drug that has the power to transform them into different people.

Perhaps this chemical, whatever it is, accounts for much of the silliness and extreme behaviour that accompany bubbles, making them unfold much like a midsummer night's dream, with people losing themselves in ill-fated delusions, mixed identities and swapped partners, until the cold light of dawn brings the world back into focus and the laws of nature and morality reassert themselves. After the dot.com bubble burst, traders were like revellers with a hangover, heads cradled in hands, stunned that they could have blown their savings on such ridiculous schemes. The shocked disbelief that the reality sustaining them for so long had turned out to be an illusion has nowhere been better described than on the front page of the *New York Times* the day after the Great Crash of 1929: 'Wall St.,' it reported, 'was a street of vanished hopes, of curiously silent apprehension and a sort of paralyzed hypnosis.'

IS THERE AN IRRATIONAL EXUBERANCE
MOLECULE?

As I say, the overconfident behaviour I describe is one that most traders will recognise and most have experienced at one point or another in their careers. I should add, however, that in addition to the changed behaviour among traders, another remarkable fact struck me during the dot.com years – that women were relatively immune to the frenzy surrounding internet and high-tech stocks. In fact, most of the women I knew, both on Wall Street and off, were quite cynical about the excitement, and as a result were often dismissed as 'not getting it', or worse, resented as perennial killjoys.

I have a special reason for relating these stories of Wall Street excess. I am not presenting them as items of front-line reportage, but rather as overlooked pieces of scientific data. Scientific research often begins with fieldwork. Fieldwork turns up curious phenomena or observations that prove to be anomalies for existing theory. The behaviour I am describing constitutes precisely this sort of field data for economics, yet it is rarely recognised as such. Indeed, out of all the research devoted to explaining financial market instability, very little has involved looking at what happens physiologically to traders when caught up in a bubble or a crash. This is an extraordinary omission, comparable to studying animal behaviour without looking at an animal in the wild, or practising medicine without ever looking at a patient. I am, however, convinced we should be looking at traders' biology. I think we should take seriously the possibility that the extreme overconfidence and risk-taking displayed by traders during a bubble may be pathological behaviour calling for biological, even clinical, study.

The 1990s were a decade ripe for such research. They gave us the folly of the dot.com bubble as well as the phrase that best described it – 'irrational exuberance'. This term, first used by Alan Greenspan in a speech delivered in Washington in 1996 and subsequently given wide currency by the Yale economist Robert Shiller, means much the same thing as an older one, 'animal spirits', coined in the 1930s by Keynes when he gestured towards some ill-defined and non-rational force animating entrepreneurial and investor risk-taking. But what are animal spirits? What is exuberance?

In the nineties, one or two people did suggest that irrational exuberance might be driven by a chemical. In 1999 Randolph Nesse, a psychiatrist at the University of Michigan, bravely speculated that the dot.com bubble differed from previous ones because the brains of many traders and investors had changed – they were under the influence of now widely prescribed antidepressant drugs, such as Prozac. 'Human nature has always given rise to booms and bubbles followed by crashes and depressions,' he argued. 'But if investor caution is being inhibited by psychotropic drugs, bubbles could grow larger than usual before they pop, with potentially catastrophic economic and political consequences.' Other observers of Wall Street, following a similar line of thought, pointed the finger at another culprit: the increasing use of cocaine among bankers.

These rumours of cocaine abuse, at least among traders and asset managers, were mostly exaggerated. (Members of the sales force, especially the salesmen responsible for taking clients out to lap-dancing bars till the wee hours of the morning, may have been another matter.) As for Nesse, his comments received some humorous coverage in the media, and when he spoke at a conference organised by the New York Academy of Sciences a year later he seemed to regret making them. But I thought he was on the right track; and to me his suggestion pointed to another possibility – that traders' bodies were producing a chemical, apparently narcotic, that was causing their manic behaviour. What was this bull-market molecule?

I came across a likely suspect purely by chance. During the later years of the dot.com era I was fortunate enough to observe some fascinating research being conducted in a neuroscience lab at Rockefeller University, a research institution hidden on the Upper East Side of Manhattan, where a friend, Linda Wilbrecht, was doing a Ph.D. I was not at Rockefeller in any formal capacity, but when the markets were slow I would jump in a taxi and run up to the lab to observe the experiments taking place, or to listen to afternoon lectures in Caspary Auditorium, a geodesic dome set in the middle of that vine-clad campus. Scientists in Linda's lab were working on what is called 'neurogenesis', the growth of new neurons. Understanding neurogenesis is in some ways the Holy Grail of the brain sciences, for if neurologists could figure out how to regenerate neurons they could perhaps cure or reverse the damage of neuro-degenerative diseases such as Alzheimer's and Parkinson's. Many of the breakthroughs in the study of neurogenesis have taken place at Rockefeller.

There was another area of the neurosciences where Rockefeller had made a historic contribution, and that was in research on hormones, and specifically their effects on the brain. Many of the breakthroughs in this field had been made by scientists addressing very specific issues in neuroscience, but today their results may help us understand irrational exuberance, for the bull-market molecule may in fact be a hormone. And if that is the case, then by a delightful coincidence, at the very moment in the late 1990s when Wall Street was asking the question 'What is irrational exuberance?', uptown at Rockefeller scientists were working on the answer.

So what exactly are hormones? Hormones are chemical messengers carried by the blood from one tissue in the body to another. We have dozens of them. We have hormones that stimulate hunger and ones that tell us when we are sated; hormones that stimulate thirst and ones that tell us when it is slaked. Hormones play a central role in what is called our body's homeostasis, the maintenance of vital signs, like blood pressure, body temperature, glucose levels, etc., within the narrow bands needed for our continued comfort and health. Most of the physiological systems that maintain our internal chemical balance operate pre-consciously, in other words without our being aware of them. For instance, we are all blissfully unaware of the Swiss-watch-like workings of the system controlling the potassium levels in our blood.

But sometimes we cannot maintain our internal balance through these silent, purely chemical reactions. Sometimes we need behaviour; sometimes we have to engage in some sort of physical activity in order to re-establish homeostasis. When glucose levels in our blood fall, for example, our bodies silently liberate glucose deposits from the liver. Soon, however, the glucose reserves burn off, and the low blood sugar communicates itself to our consciousness by means of hunger, a hormonal signal that spurs us to search for food and then to eat. Hunger, thirst, pain, oxygen debt, sodium hunger and the sensations of heat and cold, for example, have accordingly been called 'homeostatic emotions'. They are called emotions because they are signals from the body that convey more than mere information – they also carry a motivation to do something.

It is enlightening to see our behaviour as an elaborate mechanism designed to maintain homeostasis. However, before we go too far down the path of biological reductionism, I have to point out that hormones do not cause our behaviour. They act more like lobby groups, recommending and pressuring us into certain types of activity. Take the example of ghrelin, one of the hormones regulating hunger and feeding. Produced by cells in the lining of your stomach, ghrelin molecules carry a message to your brain saying in effect, 'On behalf of your stomach we urge you to eat.' But your brain does not have to comply. If you are on a diet, or a religious fast, or a hunger strike, you can choose to ignore the message. You can, in other words, choose your actions, and ultimately you take responsibility for them. Nonetheless, with the passing of time the message, at first whispered, becomes more like a foghorned bellow, and can be very hard to resist. So when we look at the effects of hormones on behaviour and on risk-taking – especially financial risk-taking – we will not be contemplating anything like biological determinism. We will be engaged rather in a frank discussion of the pressures, sometimes very powerful, these chemicals bring to bear on us during extreme moments in our lives.

One group of hormones has particularly potent effects on our behaviour – steroid hormones. This group includes testosterone, oestrogen and cortisol, the main hormone of the stress response. Steroids exert particularly widespread effects because they have receptors in almost every cell in our body and brain. Yet it was not until the 1990s that scientists began to understand just how these hormones influence our thinking and behaviour. Much of the work that led to this understanding was conducted in the lab of Bruce McEwen, a renowned professor at Rockefeller. He and his colleagues, including Donald Pfaff and Jay Weiss, were among the first scientists not only to map steroid receptors in the brain but also to study how steroids affect the structure of the brain and the way it works.

Before McEwen began his research, scientists widely believed that hormones and the brain worked in the following way: the hypothalamus, the region of the brain controlling hormones, sends a signal through the blood to the glands producing steroid hormones, be they testes, ovaries or adrenal glands, telling them to increase hormone production. The hormones are then injected into the blood, fan out across the body, and exert their intended effects on tissues such as heart, kidneys, lungs, muscles, etc. They also make their way back to the hypothalamus itself, which senses the higher hormone levels and in response tells the glands to stop producing the hormone. The feedback between hypothalamus and hormone-producing gland works much like a thermostat in a house, which senses cold and turns on the heating, and then senses the warmth and turns it off.

McEwen and his lab found something far more intriguing. Feedback between glands and the hypothalamus does indeed exist, is one of our most important homeostatic mechanisms, but McEwen discovered that there are steroid receptors in brain regions *other* than the hypothalamus. McEwen's model of hormones and the brain works in the following way: the hypothalamus sends a message to a gland instructing it to produce a hormone; the hormone fans out across the body, having its physical effects, but it also returns to the brain, changing the very way we think and behave. Now, that is one potent chemical. Indeed, subsequent research by McEwen and others showed that a steroid hormone, because of its widespread receptors, can alter almost every function of our body (its growth, shape, metabolism, immune function) and of our brain (its mood and memory) and of our behaviour.

McEwen's research was a landmark achievement because it showed how a signal from our body can change the very thoughts we think. And it raised a series of questions that today lie at the heart of our understanding of body and brain. Why does the brain send a signal to the body telling it to produce a chemical which in turn changes the way the brain works? What a strange thing to do. If the brain wants to change the way it thinks, why not keep all the signalling within the brain? Why take such a roundabout route through the body?

And why would a single molecule, like a steroid, be entrusted with such a broad mandate, simultaneously changing both body and brain? I think the answer to these questions goes something like this: steroid hormones evolved to coordinate body, brain and behaviour during archetypal situations, such as fighting, fleeing, feeding, hunting, mating and struggling for status. At important moments like these you need all your tissues cooperating on the task at hand; you do not want to be multi-tasking. It would make little sense to have, say, a cardiovascular system geared up for a fight, a digestive system primed for ingesting a turkey dinner, and a brain in the mood for wandering through fields of daffodils. Steroids, like a drill sergeant, ensure that body and brain fall into line as a single functioning unit.

The ancient Greeks believed that at archetypal moments in our lives we are visited by the gods, that we can feel their presence because these moments – of battle, of love, of childbearing – are especially vivid, are remembered as defining moments in our lives, and during them we seem to enjoy special powers. But alas, it is not one of the Olympian gods, poor creatures of abandoned belief that they are, who touches us at these moments: it is one of our hormones.

During moments of risk-taking, competition and triumph, of exuberance, there is one steroid in particular that makes its presence felt and guides our actions – testosterone. At Rockefeller University I came across a model of testosterone-fuelled behaviour that offered a tantalising explanation of trader behaviour during market bubbles, a model taken from animal behaviour called 'the winner effect'.

In this model, two males enter a fight for turf or a contest for a mate and, in anticipation of the competition, experience a surge in testosterone, a chemical bracer that increases their blood's capacity to carry oxygen and, in time, their lean-muscle mass. Testosterone also affects the brain, where it increases the animal's confidence and appetite for risk. After the battle has been decided the winner emerges with even higher levels of testosterone, the loser with lower levels. The winner, if he proceeds to a next round of competition, does so with already elevated testosterone, and this androgenic priming gives him an edge, helping him win yet again. Scientists have replicated these experiments with athletes, and believe the testosterone feedback loop may explain winning and losing streaks in sports. However, at some point in this winning streak the elevated steroids begin to have the opposite effect on success and survival. Animals experiencing this upward spiral of testosterone and victory have been found after a while to start more fights and to spend more time out in the open, and as a result they suffer an increased mortality. As testosterone levels rise, confidence and risk-taking segue into overconfidence and reckless behaviour.

Could this upward surge of testosterone, cockiness and risky behaviour also occur in the financial markets? This model seemed to describe perfectly how traders behaved as the bull market of the nineties morphed into the tech bubble. When traders, most of whom are young males, make money, their testosterone levels rise, increasing their confidence and appetite for risk, until the extended winning streak of a bull market causes them to become every bit as delusional, overconfident and risk-seeking as those animals venturing into the open, oblivious to all danger. The winner effect seemed to me a plausible explanation for the chemical hit traders receive, one that exaggerates a bull market and turns it into a bubble. The role of testosterone could also explain why women seemed relatively unaffected by the bubble, for they have about 10 to 20 per cent of the testosterone levels of men.

During the dot.com bubble, when considering this possibility, I was particularly swayed by descriptions of the mood-enhancing effects of testosterone voiced by people who had been prescribed it.

Patients with cancer, for example, are often given testosterone because, as an anabolic steroid – one that builds up energy stores such as muscle – it helps them put on weight. One brilliant and particularly influential description of its effects was written by Andrew Sullivan and published in the *New York Times Magazine* in April 2000. He vividly described injecting a golden, oily substance about three inches into his hip, every two weeks: 'I can actually feel its power on almost a daily basis,' he reported. 'Within hours, and at most a day, I feel a deep surge of energy. It is less edgy than a double espresso, but just as powerful. My attention span shortens. In the two or three days after my shot, I find it harder to concentrate on writing and feel the need to exercise more. My wit is quicker, my mind faster, but my judgment is more impulsive. It is not unlike the kind of rush I get before talking in front of a large audience, or going on a first date, or getting on an airplane, but it suffuses me in a less abrupt and more consistent way. In a word, I feel braced. For what? It scarcely seems to matter.' Sullivan could just as easily have been describing what it feels like to be a trader on a roll.

IRRATIONAL PESSIMISM

If testosterone seemed a likely candidate for the molecule of irrational exuberance, another steroid seemed a likely one for the molecule of irrational pessimism – cortisol.

Cortisol is the main hormone of the stress response, a bodywide response to injury or threat. Cortisol works in tandem with adrenalin, but while adrenalin is a fast-acting hormone, taking effect in seconds and having a half-life in the blood of only two to three minutes, cortisol kicks in to support us during a long siege. If you are hiking in the woods and hear a rustle in the bushes, you may suspect the presence of a grizzly bear, so the shot of adrenalin you receive is designed to carry you clear of danger. If the noise turns out to be nothing but wind in the leaves you settle down, and the adrenalin quickly dissipates. But if you are in fact being stalked by a predator and the chase lasts several hours, then cortisol takes over the management of your body. It orders all long-term and metabolically expensive functions of the body, such as digestion, reproduction, growth, storage of energy, and after a while even immune function, to stop. At the same time, it begins to break down energy stores and flush the liberated glucose into your blood. In short, cortisol has one main and far-reaching command: glucose now! At this crucial moment in your life, cortisol has in effect ordered a complete retooling of your body's factories, away from leisure and consumption goods to war matériel.

In the brain, cortisol, like testosterone, initially has the beneficial effects of increasing arousal and sharpening attention, even promoting a slight thrill from the challenge, but as levels of the hormone rise and stay elevated, it comes to have opposite effects – the difference between short-term and long-term exposure to a hormone is an important distinction we will look at in this book – promoting feelings of anxiety, a selective recall of disturbing memories, and a tendency to find danger where none exists. Chronic stress and highly elevated stress hormones among traders and asset managers may thus foster a thorough and perhaps irrational risk-aversion.

The research I encountered on steroid hormones thus suggested to me the following hypothesis: testosterone, as predicted by the winner effect, is likely to rise in a bull market, increase risk-taking, and exaggerate the rally, morphing it into a bubble. Cortisol, on the other hand, is likely to rise in a bear market, make traders dramatically and perhaps irrationally risk-averse, and exaggerate the sell-off, morphing it into a crash. Steroid hormones building up in the bodies of traders and investors may thus shift risk preferences systematically across the business cycle, destabilising it.

If this hypothesis of steroid feedback loops is correct, then to understand how financial markets function we need to draw on more than economics and psychology; we need to draw as well on medical research. We need to take seriously the possibility that during bubbles and crashes the financial community, suffering from chronically elevated steroid levels, may develop into a clinical population. And that possibility profoundly changes the way we see the markets, and the way we think about curing their pathologies.

In time, and with the encouragement of several colleagues, I concluded that this hypothesis should be tested. So I retired from Wall Street and returned to the University of Cambridge, where I had previously completed a Ph.D in economics. I spent the next four years retraining in neuroscience and endocrinology, and began designing an experimental protocol to test the hypothesis that the winner effect exists in the financial markets. I then set up a series of studies on a trading floor in the City of London. The results from these experiments provided solid preliminary data supporting the hypothesis that hormones, and signals from the body more generally, influence the risk-taking of traders. We will look at these results later in the book.

MIND AND BODY IN THE FINANCIAL MARKETS

Research on body–brain feedback, even within physiology and neuroscience, is relatively new, and has made only limited inroads into economics. Why is this? Why have we for so long ignored the fact that we have bodies, and that our bodies affect the way we think?

The most likely reason is that our thinking about the mind, the brain and behaviour has been moulded by a powerful philosophical idea we inherited from our culture – that of a categorical divide between mind and body. This ancient notion runs deep in the Western tradition, channelling the riverbed along which all discussion of mind and body has flowed for almost 2,500 years. It originated with the philosopher Pythagoras, who needed the idea of an immortal soul for his doctrine of reincarnation, but the idea of a mind–body split was cast in its most durable form by Plato, who claimed that within our decaying flesh there flickers a spark of divinity, this being an eternal and rational soul. The idea was subsequently taken up by St Paul and enthroned as Christian dogma. It was by that very edict also enthroned as a philosophical conundrum later known as the mind–body problem; and later physicists such as René Descartes, a devout Catholic and committed scientist, wrestled with the problem of how this disembodied mind could interact with a physical body, eventually coming up with the memorable image of a ghost in the machine, watching and giving orders.

Today Platonic dualism, as the doctrine is called, is widely disputed within philosophy and mostly ignored in neuroscience. But there is one unlikely place where a vision of the rational mind as pure as anything contemplated by Plato or Descartes still lingers – and that is in economics.

Many economists, or at any rate those adhering to a widely adopted approach known as neo-classical economics, assume our behaviour is volitional – in other words, we choose our course of behaviour after thinking it through – and guided by a rational mind. According to this school of thought, we are walking computers who can calculate the rewards of each course of action open to us at any given moment, and weight these rewards by the probability of their occurrence. Behind every decision to eat sushi or pasta, to work in aeronautics or banking, to invest in General Electric or Treasury bonds, there purr the opti-mising calculations of a mainframe computer.

The economists making these claims recognise that most people regularly fall short of this ideal, but justify their austere assumption of rationality by claiming that people behave, on average, 'as if' they had performed the actual calculations. These economists also claim that any irrationality we display in our personal lives tends to fall away when we have to deal with something as important as money; for then we are at our most cunning, and come pretty close to behaving as predicted by their models. Besides, they add, if we do not act ration-ally with our money we will be driven to bankruptcy, leaving the market in the hands of the truly rational. That means economists can continue studying the market with an underlying assumption of rationality.

This economic model is ingenious, at moments quite beautiful, and for good reason has wielded enormous influence on generations of economists, central bankers and policy-makers. Yet despite its elegance, neo-classical economics has come under increasing criti-cism from experimentally-minded social scientists who have patiently catalogued the myriad ways in which decisions and behaviours of both amateur and professional investors stray from the axioms of rational choice. One reason for its lack of realism is, I believe, that neo-classical economics shares a fundamental assumption with Platonism – that economics should focus on the mind and the thoughts of a purely rational person. Consequently, neo-classical economics has largely ignored the body. It is economics from the neck up.

What I am saying is that something very like the Platonic mind–body split lingers in economics, that it has impaired our ability to understand the financial markets. If we want to understand how people make financial decisions, how traders and investors react to volatile markets, even how markets tend to overshoot sensible levels, we need to recognise that our bodies have a say in our risk-taking. Many economists might reiterate that the importance of money ensures that we act rationally where it is concerned; but perhaps it is this very importance which guarantees a powerful bodily response. Money may be the last thing about which we can remain cool.

Economics is a powerful theoretical science, with a growing body of experimental results. In fact many economists have come to question the assumption of a Spock-like rationality, even as a simplifying assumption, and a noteworthy group among them, beginning with the Chicago economist Richard Thaler and two psychologists, Daniel Kahneman and Amos Tversky, have started a rival school known as behavioural economics. Behavioural economists have succeeded in building up a more realistic picture of how we behave when dealing with money. But their important experimental work could today easily extend to the physiology underlying economic behaviour. And signs are some economists are heading that way. Daniel Kahneman, for one, has conducted research in the physiology of attention and arousal, and has recently pointed out that we think with our body.

He is right. We do. To understand just how our body affects our brain we should first recognise that they evolved together to help us physically pursue an opportunity or run away from a threat. When confronted by an opportunity for gain, such as food or territory or a bull market, or a threat to our well-being, such as a predator or a bear market, our brain sparks a storm of electrical activity in our skeletal muscles and visceral organs, and precipitates a flood of hormones throughout our bodies, altering metabolism and cardiovascular function in order to sustain a physical response. These somatic and visceral signals then feed back on the brain, biasing our thinking – our attention, mood, memory – so that it is in sync with the physical task at hand. In fact, it may be more scientifically accurate, although semantically difficult, to stop speaking in terms of brain and body at all, as if they were separable, and to speak instead of a whole-person response to events.

Were we to start viewing ourselves in this manner we would find economics and the natural sciences beginning to merge. Such a prospect may seem futuristic and strike some people as scary and a touch dehumanising. Scientific progress, admittedly, often heralds an ugly new world, divorced from traditional values, dragging us in a direction we do not want to go. But occasionally science does not do that; occasionally it merely reminds us of something we once knew, but have forgotten. That would be the case here. For the type of economics suggested by recent advances in neuroscience and physiology merely points us back to an ancient, commonsensical and reassuring tradition in Western thought, but one that has been buried under archaeological layers of later ideas – and that is the type of thinking begun by Aristotle. For Aristotle was the first and one of the greatest biologists, perhaps the closest and most encyclopaedic observer of the human condition, and for him, unlike Plato, there was no mind–body split.

In his ethical and political works Aristotle tried to bring thought down to earth, the catchphrase of the Aristotelians being 'Think mortal thoughts'; and he based his political and ethical thinking on the behaviour of actual humans, not idealised ones. Rather than wagging a finger at us and making us feel shame for our desires and needs and the great gap existing between our actual behaviour and a life of pure reason, he accepted the way we are. His more humane approach to understanding behaviour is today in the process of being rediscovered. In Aristotle we have an ancient blueprint of how to

Fig. 1. Detail from Raphael's *School of Athens*. Plato, on the left, holds
a copy of his dialogue the *Timaeus* and points to the heavens.
Aristotle holds a copy of his *Ethics* and gestures to the world around
him, although with the palm of his hand facing down he also seems
to be saying, 'Plato, my friend, keep your feet on the ground.'

merge nature and nurture, how to design institutions so that they accommodate our biology.

Economics in particular could benefit from this approach, for economics needs to put the body back into the economy. Rather than assuming rationality and an efficient market – the unfortunate upshot of which has been a trading community gone feral – we should study the behaviour of actual traders and investors, much as the behavioural economists do, only we should include in that study the influence of their biology. If it turns out that their biology does indeed exaggerate bull and bear markets then we have to think anew about how to alter training programmes, management practices, even government policies in order to counteract it.

At the moment, though, I fear we have the worst of both worlds – an unstable biology coupled with risk-management practices that increase risk limits during the bubble and decrease them during the crash, plus a bonus scheme that rewards high-variance trading. Today nature and nurture conspire in creating recurrent disasters. More effective policies will have to consider ways of managing the biology of the market. One way to do that may be to encourage a more even balance within the banks between men and women, young and old, for each has a very different biology.

WHAT UNITES US

To begin the story I want to tell, we need to get a better understanding of how brain and body cooperate in producing our thoughts and behaviour, and ultimately our risk-taking. The best way to do that is to look at what might be called the central operation of our brain. What might that be? We may be tempted to answer, given our heritage, that the central, most defining feature of our brain is its capacity for pure thought. But neuroscientists have discovered that conscious, rational thought is a bit player in the drama that is our mental life. Many of these scientists now believe that we are getting closer to the truth if we say that the basic operation of the brain is the organisation of movement.

That statement may come as something of a shock – I know it did for me – even a disappointment. But had I learned its truth earlier than I did, I would have saved myself years of misunderstanding. You see, it is common when starting out in neuroscience to go looking for the computer in the brain, for our awesome reasoning capacities; but if you approach the brain with that goal you inevitably end up disappointed, for what you find is something a lot messier than expected. For the brain regions processing our reasoning skills are inextricably tangled up with motor circuits. You tend to get a bit annoyed at the lack of simplicity in this architecture, and frustrated at the inability to isolate pure thought. But that frustration comes from starting out with the wrong set of assumptions.

If, however, you view your brain and body and behaviour with a robust appreciation of the fact that you are built to move, and if you let that simple fact sink in, then I am willing to bet you will never see yourself in quite the same way again. You will come to understand why you feel so many of the things you do, why your reactions are often so fast as to leave conscious thought behind, why you rely on gut feelings, why it is that during the most powerful moments of your life – satisfying moments of flow, of insight, of love, of risk-taking, and traumatic moments of fear, anger and stress – you lose any awareness of a split between mind and body, and they merge as one. Seeing yourself as an inseparable unity of body and brain may involve a shift in your self-understanding, but it is, I believe, a truly liberating one.

story that is most prone to misunderstanding. It can too easily imply that our bodies became ever less important to our success as a species. An extreme example of this view can be found in science fiction, where future humans are frequently portrayed as all head, a bulbous cranium sitting atop an atrophied body. Bodies, in sci-fi and to a certain extent in the popular imagination, are seen as relics of a bestial prehistory best forgotten.

The very existence of such a story, lurking in the popular imagination, is yet another testament to the staying power of the ancient notion of a mind–body split, according to which our bodies play a secondary and largely mischievous role in our lives, tempting us from the path of reason. Needless to say, such a story is simplistic. Body and brain evolved together, not separately. Some scientists have recently begun to study the ways in which the lines of communication between body and brain became more elaborate in humans compared to other animals, how over time the brain became *more* tightly bound to the body, not less. With the benefit of their research we can discern another story about our history that is at once more complete and far more intriguing – that the true miracle of human evolution was the development of advanced control systems for synchronising body and brain.

In modern humans the body and brain exchange a torrent of information. And the exchange takes place between equals. We tend to think it does not, that information from the body constitutes nothing more than mere data being input into the computer in our head, the brain then sending back orders on what to do. The brain as puppet master, the body as puppet, to change analogies. But this picture is all wrong. The information sent by the body registers as a lot more than mere data; it comes freighted with suggestions, sometimes merely whispered, at others forcefully shouted, on how your brain should use it. You experience the more insistent of these informational prods as desires and emotions, the more subtle and dimly discernible as gut feelings. Over the long years of our evolutionary prehistory, this bodily input to our thinking has proved essential for fast actions and good judgement. Indeed, if we take a closer look at the dialogue between body and brain we will come to appreciate just how crucially

36

the body contributes to our decision-making, and especially to our risk-taking, even in the financial markets.

WHY ANIMALS CAN'T PLAY SPORTS

To free ourselves from the philosophical baggage that has impeded our understanding of body and brain, we should begin by asking a very basic question, perhaps the most basic in all the neurosciences: why do we have a brain? Why do some living creatures, like animals, have a brain, while others, like plants, do not?

Daniel Wolpert, an engineer and neuroscientist at the University of Cambridge, provides an intriguing answer to this question when he tells the tale of a distant cousin of humans, a sea squirt called the tunicate. The tunicate is born with a small brain, called a cerebral ganglion, complete with an eyespot for sensing light, and an otolith, a primitive organ which senses gravity and permits the tunicate to orient itself horizontally or vertically. In its larval stage the tunicate swims freely about the sea searching for rich feeding grounds. When it finds a promising spot it cements itself, head-first, to the sea floor.

Fig. 2. The bluebell tunicate.

It then proceeds to ingest its brain, using the nutrients to build its siphons and tunic-like body. Swaying gently in the ocean currents, filtering nutrients from passing water, the tunicate lives out its days without the need or burden of a brain.

To Wolpert, and many like-minded scientists, the tunicate is sending us an important message from our evolutionary past, telling us that if you do not need to move, you do not need a brain. The tunicate, they say, informs us that the brain is fundamentally very practical, that its main role is not to engage in pure thought but to plan and execute physical movement. What is the point, they ask, of our sensations, our memories, our cognitive abilities, if these do not lead at some point to action, be it walking, or reaching, or swimming, or eating, or even writing? If we humans did not need to move then perhaps we too would prefer to ingest our brain, a metabolically expensive organ, consuming some 20 per cent of our daily energy. Scientists who believe the brain evolved primarily to control movement – Wolpert calls himself and his colleagues 'motor chauvinists' – argue that thought itself is best understood as planning; even higher forms of thought, such as philosophy, the epitome of disembodied speculation, proceed, they argue, by hijacking algorithms originally developed to help us plan movements. Our mental life, they argue, is inescapably embodied. Andy Clark, a philosopher from Edinburgh, has put this point nicely when he states that we have inherited 'a mind on the hoof'.

To understand the brain, therefore, we need to understand movement. Yet that has turned out to be a lot harder than anyone imagined, harder in a sense than understanding the products of the intellect. We tend to believe that what belongs in the pantheon of human achievement are the books we have written, the theorems we have proved, the scientific discoveries we have made, and that our highest calling involves a turning away from the flesh, with its decay and temptation, and towards a life of the mind. But such an attitude often blinds us to the extraordinary beauty of human movement, and to its baffling mystery.

Such is the conclusion drawn by many engineers who have tried to model human movement or to replicate it with a robot. They have quickly come to a sobering realisation – that even the simplest of human movements involves a mind-boggling complexity. Steven Pinker, for example, points out that the human mind is capable of understanding quantum physics, decoding the genome and sending a rocket to the moon; but these accomplishments have turned out to be relatively simple compared to the task of reverse-engineering human movement. Take walking. A six-legged insect, even a four-legged animal, can always keep a tripod of three legs on the ground to balance itself while walking. But how does a two-legged creature like a human do it? We must support our weight, propel ourselves forward, and maintain our centre of gravity, all on the ball of a single foot. When we walk, Pinker explains, 'we repeatedly tip over and break our fall in the nick of time'. The seemingly simple act of taking a step is in truth a technical tour de force, and, he reports, 'no one has yet figured out how we do it'. If we want to observe the true genius of the human nervous system, we should therefore look not so much to the works of Shakespeare or Mozart or Einstein, but to a child building a Lego castle, or a jogger running over an uneven surface, for their movements entail solving technical problems which for the moment lie beyond the ken of human understanding.

Wolpert has come to a similar conclusion. He points out that we have been able to program a computer to beat a chess grandmaster because the task is merely a large computational problem – work out all possible moves to the end of the game and choose the best one – and can be solved by throwing a lot of computing power at it. But we have not yet been able to build a robot with the speed and manual dexterity of an eight-year-old child.

Our physical abilities are awe-inspiring, and they remain so even when compared to those of the animals. We tend to think that as we evolved out of our bodies and into our larger brains we left physical prowess behind, with the brutes. We may have a larger prefrontal cortex relative to brain size than any animal, but animals outclass us in pretty well any measure of physical performance. We are not as large as an elephant, not as strong as a gorilla, nor as fast as a cheetah. Our nose is not nearly as sensitive as a dog's, nor our eyes those of an owl. We cannot fly like a bird, nor can we swim underwater for as long as a seal. We get lost easily in the forest and end up walking in circles, while bats have radar and monarch butterflies have GPS. The gold medals for physical achievement in all events therefore go to members of the animal kingdom.

But is this true? We need to look at the question from another angle. For what is truly extraordinary about humans is our ability to learn physical movements that do not in some sense come naturally, like dancing ballet, or playing the guitar, or performing gymnastics, or piloting a plane in an aerial dogfight, and to perfect them. Consider, for example, the skills displayed by a downhill skier who, in addition to descending a mountain at over 90 miles an hour, must carve turns, sometimes on sheer ice, at just the right time, a few milliseconds separating a winning performance from a deadly accident. This is a remarkable achievement for a species that took to the slopes only recently. No animal can do anything like this. Little wonder that Olympic events draw such large crowds – we are witnessing a physical perfection unequalled in the animal world.

Remarkable feats of physical prowess can also be viewed in the concert hall. The fingers of a master pianist can disappear in a blur of movement when engaged in a challenging piece. All ten fingers work simultaneously, striking keys so rapidly that the eye cannot follow, yet each one can be hitting a key with varying force and frequency, some lingering to hold the note, others pulling back almost instantly, the whole performance modulated so as to communicate an emotional tone or conjure up a certain image. The physical feat by itself is extraordinary, but to think that this frenzied activity is so closely controlled that it can produce artistic meaning almost beggars belief. A piano concert is an extraordinary thing to watch and hear.

Humans have always dreamed of breaking the bonds of terrestrial enslavement, and in sport, as in music and dance, we have come close to succeeding. Our incomparable prowess led Shakespeare to sing of our bodies, 'In form and moving how express and admirable! In action how like an Angel!' We have to wonder, how did we develop this physical genius? How did we learn to move like the gods? We did so because we grew a larger brain. And with that larger brain came ever more subtle physical movements, and ever more dense connections with the body.

The brain region that experienced the most explosive growth in humans was, of course, the neo-cortex, home to choice and planning. The expanded neo-cortex led to the glories of higher thought; but it should be pointed out that the neo-cortex evolved together with an expanded cortico-spinal tract, the bundle of nerve fibres controlling the body's musculature. And the larger neo-cortex and related nerves permitted a new and revolutionary type of movement – the voluntary control of muscles and the learning of new behaviour. The neo-cortex did indeed give us reading, writing, philosophy and mathematics, but first it gave us the ability to learn movements we had never performed before, like making tools, throwing a spear, or riding a horse.

There was, however, another brain region which actually outgrew the neo-cortex and contributed importantly to our physical prowess – the cerebellum (see fig. 3). The cerebellum occupies the lower part of the bulge that sticks out of the back of your head. It stores memories of how to do things, like ride a bike or play the flute, as well as programmes for rapid, automatic movements. But the cerebellum is an odd part of the brain, because it seems tacked on, almost like a small, separate brain. And in some sense it is, because the cerebellum acts like an operating system for the rest of the nervous system. It makes neural operations faster and more efficient, its contribution to the brain being much like that of an extra RAM chip added to a computer. The cerebellum plays this role most notably in the motor circuits of our nervous system, for it coordinates our physical actions, gives them precision and split-second timing. When the cerebellum is impaired, as it is when we are drunk, we can still move, but our actions become slow and uncoordinated. Intriguingly, though, the cerebellum also streamlines the performance of the neo-cortex itself. In fact, there is archaeological evidence indicating that modern humans may actually have had a smaller neo-cortex than the troll-like Neanderthals; but we had a larger cerebellum, and it provided us with what was effectively a more efficient operating system, and hence more brainpower.

41

The expanded cerebellum led to our unparalleled artistic and sporting achievements. It contributed as well to the expertise we rely on when we entrust ourselves to the hands of a surgeon. Today, when our body and brain embrace, when we apply our formidable intelligence to physical action, we produce movements that are like nothing else ever seen on earth. This is a uniquely human form of excellence, and it deserves as much highbrow recognition as the works of philosophy, literature and science that occupy our pantheons.

REVVING THE BRAIN

Movement needs energy, and that means the brain has to organise not only the movement itself, but also the support operations for the muscles. What are these operations? It turns out that they are not all that different from those of an internal combustion engine. The brain must organise the finding and ingesting of fuel, in our case food; it must mix the fuel with oxygen in order to burn it; it must regulate the flow of blood in order to deliver this fuel and oxygen to cells throughout the body; it must cool this engine before combustion causes it to overheat; and it must vent the carbon dioxide waste once the fuel is burned.

These simple facts of engineering mean that our thoughts are intimately tied to our physiology. Decisions are decisions to do something, so our thoughts come freighted with physical implications. They are accompanied by a rapid shift in our motor, metabolic and cardiovascular systems as these prepare for the movements that may ensue. Thinking about the options open to us at any given moment, scrolling through the possibilities, triggers a rapid series of somatic shifts. You can often see this in a person's face as they think – eyes widening or squinting, pupils dilating, skin flushing or blanching, facial expressions as labile and fleeting as the weather. All thoughts involving choice of action involve a kaleidoscopic shift from one bodily state to another. Choice is a whole-body experience.

We are forcefully reminded of this fact whenever we contemplate the taking of risks, especially in the financial markets. When reading of the outbreak of war, for example, or watching stock prices crash, the information provokes a strong bodily response: you inhale a quick lungful of air, your stomach knots and muscles tense, your face flushes, you feel the thump, thump of a heart gearing up for action, and a thin sheen of sweat creeps across your skin. We are all so familiar with these physical effects that we take them for granted and lose sight of their significance. For the fact that information, mere letters on a page or prices on a screen, can provoke a strong bodily reaction, can even, should it create uncertainty and stress, make us physically ill, tells us something important about the way we are built. We do not regard information as a computer would, dispassionately; we react to it physically. Our body and brain rev up and down together. Indeed, it is upon this very simple piece of physiology that much of the entertainment industry is built: would we read novels or go to the movies if they did not take our bodies on a rollercoaster ride?

The point is this, and I cannot emphasise it enough: when faced by situations of novelty, uncertainty, opportunity or threat, you feel the things you do because of changes taking place in your body as it prepares for movement. Stress is a perfect illustration of this point. We tend to think that stress consists primarily of troubling thoughts, of being upset because something bad has happened or is going to happen to us, that it is a purely psychological state. But in fact the unpleasant and dangerous aspects of the stress response – the nervous stomach, the high blood pressure, the elevated glucose levels, the anxiety – should be understood as the gastro-intestinal, cardiovascular, metabolic and attentional preparation for impending physical effort. Even the gut feelings upon which traders and investors rely should be seen in this light: these are a lot more than mere hunches about what will happen next; they are changes taking place in the bodies of traders and investors as they prepare an appropriate physical response, be it fighting, running away, celebrating, or whimpering for relief. And because movement in times of emergency has to be lightning fast, these gut feelings are generated quickly, often faster than consciousness can keep up with, and are transmitted to parts of the brain of which we have only a dim and diffuse awareness.

CONTROLLING OUR INTERNAL WEATHER

For body and brain to be unified in this way, they must conduct a non-stop dialogue, a process, mentioned above, called homeostasis. Oxygen levels in the blood must be maintained within tight bands, and are kept so by a largely unconscious modulation of our breathing, as must heart rate and blood pressure. Body temperature too must be maintained within a degree or two of 37 degrees Celsius. Should it drop, say, below this band, the brain instructs our muscles to shiver and adrenal glands to raise our core temperature. Blood sugar levels too must be reported and then maintained within narrow bands, and should they fall, bringing on symptoms of low blood sugar, the brain promptly responds with a number of hormones, including adrenalin and glucagon, which liberate glucose stores for release into the blood. The amount of bodily signals being processed by the brain, coming as they do from almost every tissue, every muscle and organ, is voluminous.

Much of this bodily regulation is a job allotted to the oldest part of the brain, known appropriately as the reptile brain, and specifically to a part of it called the brain stem (see fig. 3). Sitting on top of the spine and looking like a small, gnarled fist, the brain stem controls many of the automatic reflexes of the body – breathing, blood pressure, heart rate, sweating, blinking, startle – plus the pattern generators that produce unthinking repetitive movements like chewing, swallowing, walking, etc. The brain stem acts as the life-support system of the body; other, more developed parts of the brain, ones responsible for, say, consciousness, can be damaged, leaving us 'brain dead', as they say, yet we can live on in a coma as along as the brain stem continues to operate. However, as animals evolved, the nervous circuitry linking their visceral organs such as the gut and the heart to the brain became more sophisticated. From amphibians and reptiles through mammals, primates and humans, the brain grew more complex, and with it came an expanded capacity for regulating the body.

An amphibian such as a frog cannot prevent the uncontrolled evaporation of water from its skin, so it must remain in or close to water at all times. Reptiles can retain water, and therefore can live in

44

both water and desert. But they, like amphibians, are cold-blooded, and that means they depend on the sun and warm rocks for their heat, and become all but immobile in cool weather. Because they do not take responsibility for controlling their body temperature, amphibians and reptiles have relatively simple brains.

Mammals, on the other hand, took on far greater control of their bodies, and therefore needed more brainpower. Most notably, they began to control their internal temperature, a process called thermoregulation. Thermoregulation is metabolically expensive, requiring mammals to burn a lot of fuel to generate body heat, to shiver when cold and sweat when hot, and to grow fur in autumn and moult in the spring. An idling mammal burns about five to ten times the energy of an idling reptile, so it needs to store a lot more fuel. As a result mammals had to develop greatly increased metabolic reserves; but once equipped with them they were free to hunt far and wide. The advent of mammals revolutionised life in the wild, and could be likened to the terrifying invention of mechanised warfare. Mammals, like tanks, could move a lot farther and a lot faster than their more primitive foes, so they proved unstoppable. But their mobility required more carefully managed supply lines, something that was accomplished by more advanced homeostatic circuitry.

Humans in turn took on even more control over their bodies than lower mammals. This development is reflected in a more advanced nervous system and a more extensive and animated dialogue between body and brain. We find some evidence for this process in studies comparing the brain structures among animals and humans. In one noteworthy study of comparative brain anatomy, a group of scientists looked at differences in the size of various brain regions (size is measured as a percentage of total brain weight) among existing primates to see which regions correlated with life span, a measure they took as a proxy for survivability. Their study showed that in addition to the neo-cortex and cerebellum, two other brain regions grew relatively larger in humans, most notably two regions playing a role in the homeostatic control of the body – the hypothalamus and the amygdala (fig. 3).

The hypothalamus, a brain region found by projecting lines in from the bridge of your nose and sideways from the front of your ears, regulates our hormones, and through them our eating, sleeping, sodium levels, water retention, reproduction, aggression and so on. It acts as the main integration site for emotional behaviour; in other words it coordinates the hormones and the brain stem and the

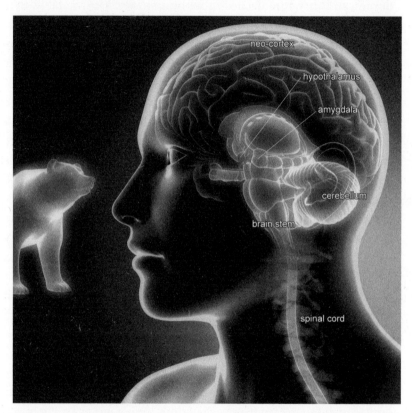

Fig. 3. Basic brain anatomy. The brain stem, often called the reptile brain, controls automatic processes such as breathing, heart rate, blood pressure, etc. The cerebellum stores physical skills and fast behavioural reactions; it also contributes to dexterity, balance and coordination. The hypothalamus controls hormones and coordinates electrical and chemical elements of homeostasis. The amygdala processes information for emotional meaning. The neo-cortex, the latest evolved layer of the brain, processes discursive thought, planning and voluntary movement. The insula (located on the far side and near the top of the illuminated brain) gathers information from the body and assembles it into a sense of our embodied existence.

emotional behaviours into a coherent bodily response. When, for example, an angry cat hisses, and arches its back, and fluffs its fur, and secretes adrenalin, it is the hypothalamus that has assembled these separate displays of anger and orchestrated them into a single coherent emotional act.

The amygdala assigns emotional significance to events. Without the amygdala, we would view the world as a collection of uninteresting objects. A charging grizzly bear would impress us as nothing more threatening than a large, moving object. Bring the amygdala online, and miraculously the grizzly morphs into a terrifying and deadly predator and we scramble up the nearest tree. The amygdala is the key brain region registering danger in the outside world and initiating the suite of physical changes known as the 'stress response'. It also registers signs of danger inside the body, such as rapid breathing and heart rate, increased blood pressure, etc., and these too can trigger an emotional reaction. The amygdala senses danger and rouses the body to high alert, and is in turn alarmed by our body's arousal, this reciprocal influence of body on amygdala, amygdala on body, occasionally feeding on itself to produce runaway anxiety and panic attacks.

Some of the most important research showing that connections between brain and body became more elaborate in humans is that conducted by Bud Craig, a physiologist at the University of Arizona. He has mapped out the nervous circuitry responsible for a remarkable phenomenon known as interoception, the perception of our inner world. We have senses like vision, hearing and smell that point outwards, to the external world; but it turns out we also have something very like sense organs that point inwards, perceiving internal organs such as the heart, lungs, liver, etc. The brain, being incurably nosey, has these listening devices – receptors that sense pain, temperature, chemical gradients, stretching tissue, immune-system activation – throughout the body, and like agents in the field they report back every detail of our viscera. This internal sensation can be brought to consciousness, as it is with hunger, pain, stomach and bowel distension, but most of it, like sodium levels or immune-system activation, remains largely unconscious, or inhabits the fringes of our awareness.

But it is this diffuse information, flowing in from all regions of the body, that gives us the sense of how we feel.

Interoceptive information is collected by a forest of nerves that flow back from every tissue in the body to the brain, travelling along nerves that feed into the spinal cord or along a superhighway of a nerve, called the vagus nerve, that travels up from the abdomen to the brain, collecting information from the gut, pancreas, heart and lungs. All this information is then channelled through various integration sites – regions of the brain that collect disparate individual sensations and assemble them into a unified experience – ending up in a region of the cortex called the insula, where something like an image of the overall state of the body is formed. Craig has looked at the nerves connecting body and brain in various animals, and has concluded that the pathways leading to the insula are present only in primates, and further that an awareness of the overall state of our body may be found uniquely in humans.

Lastly and most controversially, Craig, along with other scientists such as Antonio Damasio and Antoine Bechara, has suggested that gut feelings and emotions, rationality and even self-consciousness itself, should be seen as more advanced tools that emerged over the course of evolution to help us regulate our body.

As evolution progressed, body and brain entwined in an ever more intimate embrace. The brain sent out fibres to touch every tissue in the body, asserting control over heart, lungs, gut, arteries and glands, cooling us when hot, warming us when cold; and the body in turn pumped message after message back into the brain, telling of its wants and needs, and making suggestions as to how the brain should behave. In this manner, feedback between body and brain became more complex and extensive, not less so. We did not grow a larger brain just to fit it inside a withering body of the kind seen in sci-fi movies. The brain grew in order to control a more sophisticated body – a body that can handle a sword like Alexander, play the piano like Glenn Gould, control a tennis racket like John McEnroe, or perform open-brain surgery like Wilder Penfield.

Through the research surveyed here, from anatomy, physiology and neuroscience, we have today come to see the body as an *éminence grise*, standing behind the brain, effectively applying pressure at just the right point, at just the right time, to help us prepare for movement. Scientists, by small steps, are thus patiently stitching closed an ancient wound opened up between mind and body. By doing so they have helped us understand how body and brain cooperate at crucial moments in our lives, like the taking of risks, including, most certainly, financial risks.

PART II

Gut Thinking

THREE

The Speed of Thought

A WAKE-UP CALL ON THE TREASURY DESK

The trading floor we will be observing belongs to a large Wall Street investment bank, located a short walk from the Stock Exchange and the Federal Reserve. We begin our visit early on a crisp morning in March. It is just past 7 a.m., darkness still shrouds the city, street lamps burn, but already bankers trickle from subway stations at Broadway, Broad Street and Bowling Green, or step from taxis and limos in front of our bank. Women in Anne Taylor and sneakers grip coffees; men in Brooks Brothers look freshly scrubbed and combed, their eyes fixed, like an athlete's, on the day ahead.

Up on the 31st floor the elevator doors open and bankers are drawn into a yawning trading room. Almost a thousand desks line its grid-work of aisles, each one cluttered with half a dozen computer screens that will soon monitor market prices, live news feeds and risk positions. Most screens are black now, but one by one they are switched on, and the floor begins to blink with neon green, orange and red. A rising hubbub absorbs individual voices. Out the front window, across the narrow street, looms another glass office tower, so close you can almost read the newspaper lying on a desk. Out the side window, lower down, climbs a listed 1920s building, its stepped-back rooftop an Art Deco masterpiece: pillars topped with hooded figures; friezes depicting sunbursts, winged creatures and mysterious symbols the meaning of which have long since been forgotten. During idle moments bankers gaze down on this lost civilisation and feel a

53

momentary nostalgia for that more glamorous time, memories of the Jazz Age being just some of the ghosts haunting this storied street.

Settling in for the day, traders begin to call London and ask what has happened overnight. Once they have picked up the thread of the market they one by one take control of the trading books, transferring the risk to New York, where it will be monitored and traded until Tokyo comes in that evening. These traders work in three separate departments – bonds (the department is often called fixed income), currencies, and commodities, while downstairs a similar-sized trading floor houses the equity department. Each department in turn is split between traders and salespeople, the salespeople of a bank being responsible for convincing their clients – pension funds, insurance companies, mutual funds, in short, the institutions managing the savings of the world – to invest their money or execute their trades with the bank's traders. Should one of these clients decide to do so, the salesperson takes an order from them to buy or sell a security, say a Treasury bond or a block of currencies, say dollar–yen, and the order is executed by the trader in charge of making markets in this instrument.

One of these clients, DuPont Pension Fund, livens up what is turning out to be an uneventful day by calling in the only big trade of the morning. DuPont has accumulated $750 million-worth of pension contributions from its employees, and needs to invest the funds. It chooses to do so in US Treasury bonds maturing in ten years, the interest payments from which will finance retiring employees' pension benefits. It is still early in the day, only 9.30, and most markets are sleepy with inactivity, but the fund manager wants to execute this trade before the afternoon. That is when the Fed will announce its decision to raise or lower interest rates. Even though the financial community widely expects it to do nothing, the fund manager does not want to take unnecessary risks. Besides, for months now she has worried about what she considers an unsustainable bull market in stocks, and the very real possibility of a crash.

The fund manager scans her telephone keyboard for the four or five banks she prefers to deal with for Treasury bonds. Morgan Stanley sent her an insightful piece of research yesterday – maybe she should give them a shot. Goldman can be aggressive on price. Deutsche Bank entertains well, and last summer the salespeople covering her out of Europe took her to Henley Regatta. After a moment's indecision she passes on these banks, and decides instead to give her pal Esmee a shot. Hitting the direct line, she says, without the usual chitchat, 'Esmee, offer $750 million ten-year Treasuries, on the hop.'

Esmee, the salesperson, covers the speaker of her phone and yells to the trader on the Treasury desk, 'Martin, offer 750 tens, DuPont!'

The trader shoots back, 'Is this in competition?' meaning is DuPont getting prices from a number of other banks. The advantage of doing a trade in competition is that DuPont ensures it gets an aggressive price; the disadvantage is that several banks would know there is a big buyer, and this may cause prices to spike before the fund gets its bonds. However, the Treasury market is now so competitive that price transparency is no longer an issue, so on balance it is probably in DuPont's interest to keep this trade quiet. Esmee relays to Martin that the trade is 'out of comp', but adds, 'Print this trade, big boy. It's DuPont.'

Looking at his broker screens, Martin sees ten-year Treasuries quoted at 100.24–100.25, meaning that one bank, trying to buy them, is bidding a price of 100.24, while another, trying to sell, is offering them at 100.25. Traders post their prices on broker screens to avoid the tedious process of calling round all the other banks to find out which ones need to trade (in that regard a securities broker is no different from a real estate agent), and also to maintain anonymity. The offer price posted right now on this broker screen is good for about $100 million only. If Martin offers $750 million to DuPont at the offered price of 100.25, he has no guarantee of buying the other $650 million at the price he sold them.

To decide on the right price, Martin must rely on his feel for the market – how deep it is, in other words how much he can buy without moving prices, and whether the market is going up or down. If the market feels strong and the offers are thinning out, he may need to offer the bonds higher than indicated on the screen, at say 100.26 or

100.27. If on the other hand the market feels weak, he may offer right at the offer side price of 100.25 and wait for the market to go down. Whatever his decision, it will involve taking a substantial risk. Nonetheless, all morning Martin has been unconsciously mapping the trading patterns on the screens – the highs and lows, the size traded, the speed of movement – and comparing them to ones stored in his memory. He now mentally scrolls through possible scenarios and the options open to him. With each one comes a minute and rapid shift in his body, maybe a slight tightening of his muscles, a shiver of dread, an almost imperceptible shot of excitement, until one option just feels right. Martin has a hunch, and with growing conviction believes the market will weaken.

'Offer at 100.25.'

Esmee relays the information to DuPont, and immediately shouts back to Martin, 'Done! Thanks, Martin; you're the man.'

Martin doesn't notice the stock compliment, just the 'done' part. He now finds himself in a risky position. He has sold $750 million-worth of bonds he does not own – selling a security you do not own is called 'shorting' – and needs to buy them. The market today may not seem much of a threat, languishing as it is, but this very lack of liquidity poses its own dangers: if the market is not trading actively, then a big trade can have a disproportionate effect on prices, and if he is not stealthy Martin could drive the market up. Besides, news by its very nature is unpredictable, so Martin cannot allow himself to be lulled into a sense of security. The ten-year Treasury bond, which is considered a safe haven in times of financial or political crisis, can increase in price by up to 3 per cent in a day, and if that happened now Martin would lose over $22 million.

He immediately broadcasts over the 'squawk box' – an intercom system linking all the bank's offices around the world – that he is looking to buy ten-years at 100.24. After a few minutes a night sales-man from Hong Kong comes back and says the Bank of China will sell him $150 million at 100.24. Salespeople from around the US and Canada come back with other sales, all different sizes, eventually amounting to $175 million. Martin is tempted to take the little profit he has already made and buy the rest of the bonds he needs, but now his hunch starts to pay off; the market is weakening, and more and more clients want to sell. The market starts to inch down: 100.23–24, 100.22–23, then 100.21–22. At this point he puts in the broker screen a bid of 100.215, a seemingly high bid considering the downward drift of the market. He immediately gets hit, buying $50 million from the first seller, then building up the ticket to $225 million as other sellers come in. Traders at other banks, seeing the size of the trade on the broker screen, realise there has been a large buyer and now reverse course, trying to buy bonds in front of Martin. Prices start to climb, and Martin scrambles to lift offers while he still has a profit, at higher and higher prices, first 100.23, then .24, finally buying the last of the bonds he needs at 100.26, slightly higher than where he sold them. But it is of no concern. He has bought back the bonds he shorted at 100.25 at an average price just under 100.23.

Martin has covered his bonds within 45 minutes, and made a tidy profit of $500,000. Esmee receives $250,000 in sales credit (her sales credit, a number that determines her year-end bonus, should represent that part of a trade's profit which can be attributed to the relationship she has built with her client. You can imagine the frequent arguments between sales and trading. Like cats and dogs). The sales manager comes over and thanks Martin for helping build a better relationship with an important client. The client is happy to have bought bonds at lower levels than the current market price of 100.26. Everyone is happy. A few more days like today, and everyone can start hinting to management, even this early in the year, their high expectations come bonus time. Martin strolls to the coffee room feeling invincible, with whispered comments trailing behind him: 'That guy's got balls, selling $750 million tens right on the offer side.'

This scenario describes what happens on a trading floor when things go right. And in general things do not go badly wrong on a Treasury trading desk. There are certainly bad days, even months; but the really fatal events, like a financial crisis, strike at other desks. The reason is that Treasury bonds are considered to be less risky than other assets, such as stocks, corporate bonds or mortgage-backed securities. So when the financial markets are racked by one of their periodic crises, clients rush to sell these risky assets and to buy Treasuries. Trading volume in Treasuries balloons, the bid–offer spread widens, and volatility spikes. In periods like that Martin may price billion-dollar deals several times a day, and instead of making one or two cents, he may make half a point – $5 million at a crack. A Treasury desk usually makes so much money during a crisis it helps buffer the losses made on other trading desks, ones more exposed to credit risk.

There is a further reason the Treasury desk holds a privileged position on a trading floor, and that is the unrivalled liquidity of Treasury bonds. A bond is said to be liquid if a client can buy and sell large blocks of it without paying a lot in bid–offer spread and broker commissions. In normal conditions, clients can buy a ten-year Treasury at the offer side price of, say, 100.25, and sell it immediately, should they need to, a mere one cent lower. By way of comparison, corporate bonds, ones issued by companies, commonly trade with a bid–offer spread of 10–25 cents, with some trading as wide as $1 or $2. The Treasury market is the most liquid of all bond markets, and is thus perfectly tailored for large flows and fast execution, Treasury bonds being the thoroughbreds of trading instruments.

Such a market calls forth traders with a complementary set of skills. Traders like Martin must price client trades quickly, and cover their positions nimbly, before the market moves against them. This is especially true when the markets pick up speed, for then Martin has no time to think; if he is to avoid owning bonds in a falling market or being caught short in a rising one, he must price and execute his trades with split-second timing. In this his behaviour resembles not so much that of rational economic man, weighing utilities and calculating probabilities, but a tennis player at net.

We are now going to look at Martin's trade much as an athlete's coach would, as a physical performance. We saw in the last chapter that our brain evolved to coordinate physical movements, and these, by the very nature of the world we lived in, had to be fast. If our actions had to be fast, so too did our thinking. As a result we came to rely on what are called pre-attentive processing, automatic motor responses and gut feelings. These processes travel a lot faster than conscious rationality, and help us coordinate thought and movement when time is short. We will look at some extraordinary research that demonstrates just how unaware we can be of what is really going on in our brains when we make decisions and take risks.

In this chapter we stray from the trading floor and visit other worlds where speed of reactions is crucial for survival, as it is in the wild and in war, and crucial for success, as it is in sports and trading. In the next chapter we look at gut feelings. These chapters provide the science we need, the background story, that will help us understand what we are seeing when, in later chapters, we head back onto the trading floor and watch Martin and his colleagues as they are swept up in a fast-moving market.

THE ENIGMA OF FAST REACTIONS

We evolved in a world where dangerous objects frequently hurtled at us at high speeds. A lion sprinting at 50 miles an hour from a hundred feet away will sink its teeth into our necks in just over one second, giving us very little time to run, climb a tree, string a bow, or even think about what to do. A spear launched in battle at 65 miles an hour from 30 feet away will pierce our chest in a little over 300 milliseconds (thousandths of a second), about a third of a second. As predator and projectile zero in, and our time to escape runs out, the speed of the reactions needed to survive shortens into a timeframe our conscious mind has difficulty imagining. Over millennia of prehistory, the difference between someone who lived and someone who died often came down to a few thousandths of a second in reaction time. Evolution, like qualifying heats at the Olympics, took place against the sustained ticking of a stopwatch.

Things are not that different today, in sport, for example, or war, or indeed in the financial markets. In sport we have sharpened the rules and honed the equipment to such an extent that once again, as in the jungle, we have pushed up against our biological speed limits. A baseball pitched at 90 miles an hour covers the 60 feet to home plate in about 450 milliseconds; a tennis ball served at 140 miles an hour will catch the service line in under 400 milliseconds; a penalty shot in football will cover the short 36 feet to the goal in about 290 milliseconds; and an ice hockey puck shot halfway in from the blue line will impact the goalie's mask in less than 200 milliseconds. In each of these cases, the less than half a second travel time of the projectile gives the receiving athlete about half that time to make a decision whether or not to swing the bat, or return the serve, or jump to the left or right, or reach for the puck, for the remaining time must be spent initiating the muscle or motor response.

Even these short timeframes do not capture the truly miraculous speeds frequently demanded of the human body. In table tennis, which many of us consider a leisurely pursuit, the ball when smashed travels at 70 miles an hour, yet the distance between players may be only 14 to 16 feet, giving the returning player about 160 milliseconds to react. The difference between winning and losing has been shaved to a few thousandths of a second in reaction times. Similar reaction times are found in sprinters, who are so fast off the blocks, reacting to the starting gun in a little over 120 milliseconds, with some even approaching the 100-millisecond mark, that races increasingly feature what are called silent guns. These starting pistols produce a bang which is heard from electronic speakers placed behind each runner so that they all hear the starting signal at the same time. Without these speakers the runners in the outside lanes would hear the pistol with a fatal 30-millisecond delay, that being about the time it takes the sound of the shot to reach them.

Or consider one of the most dangerous positions in the sporting world, the short distance infielder in cricket. On a cricket field, one brave soul plants himself, crouched at the ready, a mere 14 to 17 feet from the batsman, with some coming in even closer than that. Here, without the benefit of gloves, he attempts either to catch the ball as it explodes off the bat, or clear out of the way. A cricket ball, slightly larger than a baseball and much harder, rebounds off a swinging bat at speeds of up to 100 miles an hour. The fielder facing this ball must first take care not to be hit by the bat itself, and then has as little as 90 milliseconds, less than a tenth of a second, to react to the incoming projectile. One of the closest of these positions is appropriately called the silly point, and in here, this close to the batsman, death can occur. One Indian player, Raman Lamba, was killed by a ball to the temple while he was fielding at another position frighteningly close to the batsman.

Equally deadly projectiles, ones responsible for far more injuries, can be found in contact sports like karate and boxing, where punches have been clocked at terrifying speeds. Norman Mailer, reporting on the Rumble in the Jungle, when Muhammad Ali fought George Foreman in the Zairean capital Kinshasa in 1974, describes Ali warming up in the ring, 'whirling away once in a while to throw a kaleidoscope-dozen of punches at the air in two seconds, no more – one-Mississippi, two-Mississippi – twelve punches had gone by. Screams from the crowd at the blur of the gloves.' If Mailer's numbers

Fig. 4. Speed of reactions. Jo-Wilfried Tsonga reaching for a volley at Wimbledon, 2011. If we assume his opponent, Novak Djokovic, hit a backhand from the baseline at about 90 mph, then Tsonga had a little over 300 milliseconds to respond.

are right, one of Ali's punches would run its course from beginning to end in about 166 milliseconds, although Foreman would only have had half that time to avoid it. In fact, later, more scientific measurement timed Ali's left jab at little more than 40 milliseconds.

It should come as no surprise that athletes facing fast-moving objects like cricket balls or ice hockey pucks frequently fail to intercept them (or in boxing to avoid them). But if an athlete succeeds, say, one time out of three, as a good baseball player does when at bat, his success rate approaches that of many predators in the wild. A lion, for example, closing in on an antelope, or a wolf on a deer, catches its prey on average one time out of three. In sport, as in nature, competition has pushed reaction times right to the frontier of the biologically possible.

Unfortunately, those of us not gifted with the reaction times of an Olympic athlete are nonetheless often called upon to respond with something like their speed, especially while on the road. A driver speeding at 70 miles an hour has as little as 370 milliseconds to avoid a car 75 feet in front that has mistakenly swerved into the oncoming lane. Here a success rate of one out of three still leaves a lot of car crashes.

The speed demanded of our physical reactions, in the wild, in sports, on the road, even in the financial markets, raises troubling questions when lined up against certain findings in neuroscience. Take this curious fact, for instance: once an image hits the retina, it takes approximately 100 milliseconds – that is a full tenth of a second – before it consciously registers in the brain. Pause for a moment and contemplate that fact. You will soon find it profoundly disturbing. We tend to think, as we survey the world around us or sit in the stands of a sporting match, that we are watching a live event. But it turns out that we are not – we are watching news footage. By the time we see something, the world has already moved on.

The trouble stems from the fact that our visual system is surprisingly slow. When light hits our retina, the photons must be translated into a chemical signal, and then into an electrical signal that can be carried along nerve fibres. The electrical signal must then travel to the very back of the brain, to an area called the visual cortex, and then project forward again, along two separate pathways, one processing the identity of the objects we see, the 'what' stream, as some researchers call it, and the other processing the location and motion of the objects, the 'where' stream. These streams must then combine to form a unified image, and only then does this image emerge into conscious awareness. The whole process is a surprisingly slow one, taking, as mentioned, up to one tenth of a second. Such a delay, though brief, leaves us constantly one step behind events.

Neuroscientists have discovered another problem with the idea that we are watching the world live. An important part of this idea is the notion that our eyes objectively and continuously record the scene before us, much like a movie camera. But eyes do not operate like this. If we continuously recorded the visual information presented to us, we would waste a great deal of time (and probably suffer constant headaches) looking at blurred images as our eyes pan from one scene to another. More importantly, we would be swamped by the sheer amount of data, most of which is irrelevant to our needs. Live streaming takes up an enormous amount of bandwidth on the internet, and it does so as well in our brains. To avoid a needless drain on our attentional resources, our brain has hit upon the tactic of sampling from a visual scene, rather than filming it. Our eyes fix on a small section of our visual field, take a snapshot, then jump to another spot, take a snapshot, and quickly jump again, much like a hummingbird nervously flitting from flower to flower. We are largely unaware of this process, and do not see a blur when our eyes shift location because, remarkably, the visual system stops sending images up to consciousness while it jumps from scene to scene. Furthermore, we are unaware of these jumps and intervening blackouts because our brain weaves these images seamlessly into something that does appear much like a movie. We can perform up to five of these visual jumps per second, the minimum amount of time required for a shift in view being therefore one fifth of a second.

If we return to sports, we can see that some numbers do not add up. How can a cricketer at the silly point catch (or duck) a ball in under a tenth of a second if he is not even aware of it yet? How can he direct his attention to the ball if it takes him twice as long just to move his eyes? And when dealing with these numbers we have not even begun to consider the additional 300–400 milliseconds required for an elementary cognitive decision or inference, and the 50 milliseconds or so it takes for a motor command to be communicated by nerves to our muscles. The picture conjured by these numbers is one of an infielder frozen in the readiness stance, eyes fixed like a waxwork statue, while a projectile shudders past his immobile and fragile head.

The same questions we ask about athletes can be asked, and with more urgency, beyond the sport field. How can we humans survive in a brutal and fast-moving world if our consciousness arrives on the scene just after an event is over? This is a baffling question. But asking it allows us to see what is wrong with the notion of the brain as a central processor, taking in objective information from the senses in the manner of a camera, processing this information rationally, consciously and discursively, deciding on the appropriate and desired action, and then issuing motor commands to our muscles, be they larynx or quadriceps. Each of these steps takes time, and if we were indeed programmed to behave this way, then life as we know it would be very different. If we had to think consciously about every action we took, sporting events would become odd, slow-motion spectacles that few people would have the patience to watch. Worse, in nature and in war we would have long ago fallen prey to some quicker beast.

I, CAMERA?

It turns out that there is something wrong with each step in this supposed chain of mental events. The eye takes snapshots rather than movies; but even these snapshots are not a photographic and objective record of the outside world. All sensory information comes to us tampered with. Like the news on TV, it is filtered, warped and pre-interpreted in a way designed to catch our attention, ease comprehension and speed our reactions.

Take for instance the ways in which the brain deals with the problem of the one-tenth-of-a-second delay between viewing a moving object and becoming consciously aware of it. Such a delay puts us in constant danger, so the brain's visual circuits have devised an ingenious way of helping us. The brain anticipates the actual location of the object, and moves the visual image we end up seeing to this hypothetical new location. In other words, your visual system fast-forwards what you see.

An extraordinary idea, but how on earth could we ever prove it to be true? Neuroscientists are devilishly clever at tricking the brain into revealing its secrets, and in this case they have recorded the visual fast-forwarding by means of an experiment investigating what is called the 'flash-lag effect'. In this experiment a person is shown an object, say a blue circle, with another circle inside it, a yellow one. The small yellow circle flashes on and off, so what you see is a blue circle with a yellow circle blinking inside it. Then the blue circle with the yellow one inside starts moving around your computer screen. What you should see is a moving blue circle with a blinking yellow one inside it. But you do not. Instead you see a blue circle moving around the screen with a blinking yellow circle trailing about a quarter of an inch behind it. What is going on is this: while the blue circle is moving, your brain advances the image to its anticipated actual location, given the one-tenth-of-a-second time lag between viewing it and being aware of it. But the yellow circle, blinking on and off, cannot be anticipated, so it is not advanced. It thus appears to be left behind by the fast-forwarded blue circle.

The eye and brain perform countless other such tricks in order to speed up our understanding of the world. Our retina tends to focus on the front edge of a moving object, to help us track it. We process more information in the lower half of our visual field, because there is normally more to see on the ground than in the sky. We group objects into units of three or four in order to perceive numbers rather than count them, a process, known as subitising, that comes in handy when assessing the number of opponents in battle. We rapidly and unconsciously assume an object is alive if it moves in certain ways, regularly changing direction say, or avoiding other objects, and then pay it closer attention than we would if it was inanimate.

Our reaction times can also be speeded up by relying more on hearing than vision. That may seem counter-intuitive. Light travels faster than sound, much faster, so visual images reach our senses before sounds. However, once the sensations reach our eyes and ears, the relative speeds of the processing circuits reverse. Hearing is faster and more acute than seeing, about 25 per cent so, and responding to an auditory cue rather than a visual one can save us up to 50 milliseconds. The reason is that sound receptors in the ear are much faster and more sensitive than anything in the eye. Many athletes, such as tennis and table-tennis players, rely on the sound a ball makes on a racket or bat as much as on the sight of its trajectory. A ball hit for speed broadcasts a different sound from one sliced or spun, and this information can save a player the precious few milliseconds that separate winners from losers.

If we now add up all the time delays between an event occurring in the outside world and our perceiving it, we discover the following lovely fact. For events occurring at a distance, we see them first and hear them with a delay, as we do, for example, when seeing lightning and hearing the thunder afterwards. But for events taking place close to us, we hear them, because of our rapid auditory system and relatively slow visual one, slightly in advance of seeing them. There is, though, a point at which sights and sounds are perceived as occurring simultaneously, and that point is located about ten to fifteen metres from us, a point known as the 'horizon of simultaneity'.

Could our more rapid hearing provide traders with an edge over competitors? Right now, all price feeds onto a trading floor are visual images on a computer screen. But the technology does exist for supplying audio price feeds. These have already been supplied to blind people, and apparently they sound much like an audiocassette on fast forward. Such a feed could give traders a 40-millisecond edge. That is not much time. But who knows, it could prove decisive when hitting a bid or lifting an offer during a fast market.

Bringing a trader's hearing into play may have a further advantage. Research in experimental psychology has found that perceptual acuity and general levels of attention increase as more senses are involved. In other words, vision becomes more acute when coupled with hearing, and both become more acute when coupled with touch. The explanation ventured for these findings is that information arriving from two or more senses instead of just one increases the probability that it is reporting a real event, so our brain takes it more seriously. Many older trading floors may have inadvertently capitalised on this phenomenon, because they came equipped with an intercom to the futures exchanges, with an announcer reporting bond futures prices: 'One, two ... one, two ... three, four ... fours gone, fives lifted, size coming in at six ...' and so on. With the advent of computerised pricing services, many companies felt this voice feed was antiquated and discontinued the service. Yet by bringing in a second sense it may have been an effective way of sharpening attention and reactions among the traders.

KNOWING BEFORE KNOWING

All these *ad hoc* adjustments to the information being transmitted to your conscious brain keep you from falling hopelessly behind the world. But the brain has an even more effective way of saving you from your fatally slow consciousness. When fast reactions are demanded it cuts out consciousness altogether and relies instead on reflexes, automatic behaviour and what is called 'pre-attentive processing'. Pre-attentive processing is a type of perception, decision-making and movement initiation that occurs without any consultation with your conscious brain, and before it is even aware of what is going on.

This processing, and its importance to survival, has nowhere been better described than in the extraordinary book *All Quiet on the Western Front*, written by Erich Maria Remarque, a soldier who served in the trenches during the First World War. Remarque explains that to survive on the front soldiers had to learn very quickly to pick out from the general din the 'malicious, hardly audible buzz' of the small shells called daisy cutters, for these were the ones that killed infantry. Experienced soldiers could do this, and developed reactions that kept them alive even amid an artillery bombardment: 'At the sound of the first droning of the shells,' Remarque tells us, 'we rush back, in one part of our being, a thousand years. By the animal instinct that is awakened in us we are led and protected. It is not conscious; it is far quicker, much more sure, less fallible, than consciousness. One cannot explain it. A man is walking along without thought or heed; – suddenly he throws himself down on the ground and a storm of fragments flies harmlessly over him; – yet he cannot remember either to have heard the shell coming or to have thought of flinging himself down. But had he not abandoned himself to the impulse he would now be a heap of mangled flesh. It is this other, this second sight in us, that has thrown us to the ground and saved us, without our knowing how.'

Neuroscientists have long known that most of what goes on in the brain is pre-conscious. Compelling evidence of this fact can be found in the work of scientists who have calculated the bandwidth of human consciousness. Researchers at the University of Pennsylvania, for example, have found that the human retina transmits to the brain approximately 10 million bits of information per second, roughly the capacity of an ethernet connection; and Manfred Zimmermann, a German physiologist, has calculated that our other senses record an additional one million bits of information per second. That gives our senses a total bandwidth of 11 million bits per second. Yet of this massive flow of information no more than about 40 bits per second actually reaches consciousness. We are, in other words, conscious of only a trivial slice of all the information coming into the brain for processing.

A fascinating example of this pre-conscious processing can be found in a phenomenon known as blindsight. It became a topic first of curiosity and then of medical concern during the First World War, when medics noticed that certain soldiers who had been blinded by a bullet or shell wound to the visual cortex (but whose eyes remained intact) were nonetheless ducking their heads when an object, such as a ball, was tossed over their heads. How could these blind soldiers 'see'? They were seeing, it was later discovered, with a more primitive part of the brain. When light enters your eye its signal follows the pathways, described above, back to your visual cortex, a relatively new part of the brain. However, part of the signal also passes down through an area called the superior colliculus, which lies underneath the cortex, in the midbrain (fig. 5). The superior colliculus is an ancient nucleus (collection of cells) that was formerly used for tracking objects, like insects or fast-moving prey, so that our reptilian ancestors could, say, zap it with their tongues. Now largely layered over by evolutionarily more advanced systems, it nonetheless still works. It is not sophisticated: it cannot distinguish colour, discern shape or recognise objects, the world appearing to the superior colliculus

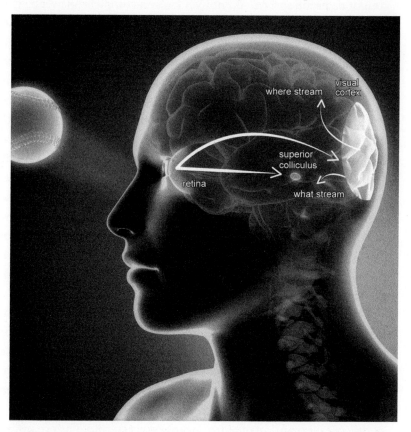

Fig. 5. The visual system. Visual images travel by electrical impulses projected from the retina to the visual cortex at the back of the brain. They are then sent forward along the 'what' stream, which identifies the object, and the 'where' stream, which identifies its location and movement. An older, faster route for visual signals travels down to the superior colliculus where fast-moving objects can be tracked.

much like an image seen through frosted glass. But it does track motion, capture attention and orient the head towards a moving object. And it is fast. Fast enough, according to some scientists, to account for a batter or infielder's rapid tracking of a moving ball. Lastly, blindsight operates without us ever being aware of it.

To what features of the world do we pre-attend? When an infielder is crouched at the ready, frozen like a statue, his eyes fixed and unable to scan, what in his visual field captures the interest of his pre-conscious processor? We do not yet know a complete answer to this question, but we do know a few things. We attend pre-consciously, as in blindsight, to moving objects, especially animate ones. We attend to images of certain primitive threats, such as snakes and spiders. And we are strongly biased to aurally attend to human voices, and visually to faces, especially ones expressing negative emotions such as fear or anger. All these objects can be registered so rapidly, in as little as 15 milliseconds (this does not include a motor response, of course), that they can affect our thinking and moods without our even being aware of them. In fact we often know whether we like or dislike something or someone well before we even know what or who it is. The speed and power of pre-conscious images, especially sexual ones, were once used in subliminal advertising as a way of biasing our subsequent spending decisions. More usefully, this pre-conscious processing can affect motor commands for reflex actions and automatic behaviours.

One of these reflexes is our startle response, a quick and involuntary contraction of muscles designed to withdraw us, like an escaping octopus, from a sudden threat. It can be initiated by both sights and sounds. A loud bang will trigger the startle, as will a rapidly approaching object in our visual field. The way we visually detect an object on a collision course with us is ingenious: our startle is initiated by a symmetrical expansion of a shadow in our visual field. The expanding shadow indicates an incoming object, and its symmetry indicates that it is heading straight for us. Apparently this pre-conscious object tracking is so well calibrated that if the shadow is expanding asymmetrically our brain can tell within five degrees that the object will miss us, and as a result the startle response is not triggered. The startle, from sensory stimulus to muscle contraction, is exceptionally fast, your head reacting in as little as 70 milliseconds and your torso, since it is farther from your brain, in about 100 milliseconds. Coincidentally, that is roughly the time required for an infielder at the silly point to catch a ball coming off a cricket bat. It is entirely possible that infielders rely on the startle response to achieve the almost inhuman response times they display. If so, then, conveniently, perhaps the infielder can catch or avoid a ball in the little time allowed him only if it is coming straight for his head.

Besides the startle response, how can we react fast enough to meet the challenges sports, and daily life, throw at us? As we saw in the previous chapter, humans have adopted a wide range of movements, like those found in sports and dance and modern warfare and even trading, for which evolution has not prepared us. How can these learned movements become so habitual that they approach the speeds needed for sporting success or survival in the wild? To answer this question we should recognise a basic principle at work in our reflexes and automatic behaviours: the higher we rise in the nervous system, moving from the spine to the brain stem to the cortex (where voluntary movement is processed), the more neurons are involved, the longer the distances covered by nervous signals, and the slower the response. To speed our reactions the brain tends therefore to pass control of the movement, once it has been learned, back to lower regions of the brain where programmes for unthinking, automatic and habitual actions are stored. Many of these learned and now-automatic behaviours can be activated in as little as 120 milliseconds.

A glimpse into this process has been provided by a brain-scanning study of people learning the computer game Tetris. At the beginning of the study, large swathes of the trainees' brains lit up, showing a complex process of learning and voluntary movement; but once they had mastered the game their movements became habitual, and brain activity in the cortex died down. Their brains now drew much less glucose and oxygen, and their speed of reactions increased markedly. Once the players had the knack, they no longer thought about playing

71

the game. This study, and others like it, supports the old saying that when learning begins we are unconscious of our incompetence, and proceed to a stage where we are conscious of our incompetence; then when training begins we move to conscious competence; and as we master our new skill we arrive at the end point of our training – unconscious competence. Thinking, one could say, is something we do only when we are no good at an activity.

One last point. As fast as these automatic reactions may be, they still do not seem quite fast enough for many of the high-speed challenges we face, and may therefore leave us slightly behind the ball, so to speak. The trouble with these reaction times is just that – they are reactions. But good athletes are not in the habit of waiting around for a ball or a fist to appear, or opponents to make their move. Good athletes anticipate. A baseball batter will study a pitcher and narrow down the likely range of his pitches; a cricket infielder will have registered a hundred tiny details of a batsman's stance and glance and grip even before the ball has left the bowler's hand; and a boxer, while dancing and parrying jabs, will pre-consciously scan his opponent's footwork and head movements, and look for the telltale setting of his stabiliser muscles as he plants himself for a knockout blow. Such information allows the receiving athlete to bring online well-rehearsed motor programmes and to prepare large muscle groups so that there is little to do while the ball or fist is in the air but make subtle adjustments based on its flightpath. Skilled anticipation is crucial to lowering reaction times throughout our physiology.

Let us finish by listening to Ken Dryden, a legendary goalie in ice hockey and one of the most articulate athletes ever, on the importance of anticipation and automatic behaviour: 'When a game gets close to me, or threatens to get close, my conscious mind goes blank. I feel nothing. I hear nothing, my eyes watch the puck, my body moves – like a goalie moves, like I move; I don't tell it to move or how to move or where, I don't know it's moving, I don't feel it move – yet it moves. And when my eyes watch the puck, I see things I don't know I'm seeing … I see something in the way a shooter holds his stick, in the way his body angles and turns, in the way he's being checked, in what he's done before that tells me what he'll do – and my body moves. I let it move. I trust it and the unconscious mind that moves it.'

To sum up, we humans have been equipped over our long evolutionary training period with a large bag of tricks designed to increase our speed of reactions. In the foregoing discussion I have rummaged in this bag and pulled out only a few of our amazing gadgets. But demonstrating how they work should be enough, I hope, to show just how reliant we are on these quick responses for survival in the wild and in war, for success in sports, and for buying back a large block of bonds sold to DuPont.

WHAT LIES BENEATH

In fact, so fast are our reactions that consciousness is frequently left out of the loop. Given that sobering fact, we have to ask: what role does consciousness play in our lives? We experience our consciousness as something residing in our heads, peering out through our eyes much as a driver peers through a windscreen, so we tend to believe that our brain interacts with our body just as a person interacts with a car, choosing the direction and speed and issuing commands to a passive and mechanical device. But this belief does not stand up to scientific scrutiny. As George Loewenstein, an economist at Yale, points out, 'There is little evidence beyond fallible introspection supporting the standard assumption of complete volitional control of behavior.' And he is right, for the stats on reaction times tell us otherwise: we are for the most part on autopilot.

The news gets even worse for the Platonists among us. In the 1970s, Benjamin Libet, a physiologist at the University of California, conducted a famous series of experiments that has tormented many a scientist and philosopher. These experiments were simplicity itself. Libet wired up a group of participants with what are called EEG leads, small monitors attached to the scalp which record the electrical activity in the brain, and then asked them to make a decision to do

73

something, like lift a finger. What he found was that the participants' brains were preparing the action 300 milliseconds before they actually made the decision to lift their finger. In other words, their conscious decision to move came almost one third of a second after their brain had initiated the movement.

Consciousness, these experiments suggested, is merely a bystander observing a decision already taken, almost like watching ourselves on video. Scientists and philosophers have proposed many interpretations of these findings, one of which is that the role of consciousness may not be so much to choose and initiate actions, but rather to observe decisions made and veto them, if need be, before they are put into effect, much as we do when we practise self-control by stifling inappropriate emotional or instinctive urges. (We may be on autopilot for much of the day, but that does not mean we cannot take responsibility for our actions.) Libet's experiments, suggesting as they do that consciousness is largely an override mechanism, led one particularly witty commentator, the Indian neuroscientist V.S. Ramachandran, to conclude that we do not in fact have free will; what we have is free won't.

It seems that consciousness is a small tip of a large iceberg. But what exactly lies below it? What lurks beneath our rational, conscious selves? The eighteenth-century German philosopher Immanuel Kant proposed a particularly intriguing answer to this question: we do not know what is down there. Kant believed that our consciousness – that is, our experience of a unified and understandable world, and of a continuing person experiencing this world – is possible only because our mind constructs this unified experience. If our mind did not organise our sensations the world would be a whirling, blooming confusion. But the mind does: it provides organising constructs, such as space and time, so that we experience a continuing world, just as it does another construct, that of cause and effect, which ties succeeding events together into a coherent story. Kant thought all these unifying constructs applied only to the veil of sensations, and not to the entities creating or lying behind the sensations. These objects we can never know. Inaccessible to rational analysis, forever mysterious to science, these hidden beings can be groped at and suggestively discerned only through art and religion. And it is in this dark world that the soul belongs, putting it too beyond the ken of rationality and beyond the domain of cause and effect. It was upon this argument that Kant rested his belief in free will.

Kant's philosophy left a deep imprint on German thought. Freud, inspired by Kant's vision, argued that below the façade of our rational selves, deep in our subconscious, there boils a devil's cauldron of envy and sexual perversion and patricidal tendencies which warps our judgement. Nietzsche too found beneath our delusions of rationality and morality a dark urge for power and dominance. Modern neuro-science, however, has lifted the lid off this hitherto mystifying brain and found something far more valuable than the entities proposed by nineteenth-century German philosophy – a meticulously engineered control mechanism. More valuable because it has been precisely cali-brated over millennia to keep us alive in a brutal and fast-moving world. And we can thank our lucky stars for it, otherwise we would long ago have been battered to extinction. Lifting the hood of our brain does not reveal the nether world of Kant's unsayable, nor the volcanic will of Nietzsche's superman, nor yet the hellish subterra-nean den of Freud's subconscious. It reveals something that is a lot closer to the inner workings of a BMW.

FAST TIMES ON THE TRADING FLOOR

Let us now return to the financial world, and consider the importance of fast reactions to the success and survival of risk-takers. Traders like Martin frequently face high-speed challenges which demand an equally fast response. The challenges may not demand quite the same speed of reactions as fielding at the silly point, but traders nonetheless regularly face time constraints, and when they do their decision-making and trade execution must bypass conscious rationality and draw instead on automatic reactions. This is especially true when markets begin to move fast, as they might in a frantic bull market. Then Martin is obliged to sell bonds to clients or risk alienating the sales force, and must scramble to buy them on the broker screens or from other clients before losing money. At times like this trading is much like a game of snap, and the fastest person wins.

This simple point carries unexpected implications for economics. It is not often appreciated that financial decision-making is a lot more than a purely cognitive activity. It is also a physical activity, and demands certain physical traits. Traders with a high IQ and insight into the value of stocks and bonds may be worth listening to, but if they do not have an appetite for risk then they will not act on their views and will suffer the fate of Cassandra, who could predict the future but could not affect its course. And even if they have a good call on the market and a healthy appetite for risk, yet are shackled with slow reactions, they will remain one step behind the market, and will not survive on the trading desk – or anywhere else in the financial world, for that matter.

Treasury traders, like flow traders more generally (a flow trader is one who trades with clients, handles the flows coming off the sales desks), therefore require a battery of traits: they need a high enough IQ and sufficient education to understand basic economics; a hearty appetite for risk; and a driving ambition. But they also need the physical build. They must be able to engage in extended periods, hours at a time, of what is called visuo-motor scanning, i.e. scrutinising the screens for price anomalies between say the ten-year and the seven-year Treasury bond, or between the bond and currency markets. Such scanning requires concentration and stamina, and not everyone can do it, just as not everyone can run a four-minute mile. And once a price discrepancy has been identified, or a high bid spotted during a sell-off, a trader must move quickly to trade on these prices before anyone else. Not surprisingly, most flow-trading desks, be they ones trading Treasury or corporate or mortgage-backed bonds, usually employ one or two former athletes, a World Cup skier, say, or a college tennis star.

The physical nature of trading is even more apparent on other types of floors. On the floor of a stock exchange or the bond and commodity pits at the Chicago Board of Trade, a trader's job can resemble a day spent in a wrestling ring. Hundreds of traders stand together, jostling each other and vying for attention when trying to trade with each other, something they do with an arcane system of hand signals. When markets are moving fast and a trader needs the attention of someone on the other side of the pit, then height, strength and speed are of paramount importance in executing a trade, as is the willingness to elbow a competitor in the face. Needless to say, there are not a lot of women in the financial mosh pits.

Another style of trading that makes punishing physical demands is what is called high-frequency trading. This activity involves buying or selling securities, say a bond or stock or futures contract, sometimes in sizes amounting to billions, but holding the positions for only a few minutes, sometimes mere seconds. High-frequency traders do not try to predict where the market is going in the next day or two, let alone the next year, as do asset managers who invest for the long term; they try to predict the small moves in the market, a few cents up or down. As a general rule, the shorter the holding period for a style of trading, the greater the need for its traders to have fast reactions.

Having said all this, there are good reasons for expecting the physical aspect of trading to decline in importance in the financial world. More and more activities are now carried out electronically. The first and most dramatic sign of such a change was the closing down of physical stock exchanges, such as the London Stock Exchange. In their place mainframe computers took over the task of matching buyers and sellers of securities. Today only a few physical exchanges, with tumultuous floors and face-to-face execution of trades, remain, the New York Stock Exchange and the Chicago Board of Trade being the most famous.

The same evolution has begun in bond and currency trading at banks. Many banks began to post the prices of the most liquid securities, beginning with Treasuries and mortgage-backed bonds, on computer screens, and then allowed their clients access to the screens. That way they could execute trades themselves, without the need of going through a salesperson like Esmee. Normally traders like Martin post prices on these screens for a limited size, say $25 to $50 million, and these will be executed electronically by clients; but for bigger trades, like DuPont's, clients still prefer to call their salesperson. Nonetheless, many people within the banks think the flow traders are dinosaurs, and will eventually go extinct.

Perhaps the greatest threat facing the human trader, though, comes from computerised trading algorithms known as black boxes. Life for many traders has always been nasty, brutish and short, given the vicious competition between them. Survival has depended on their relative endowment of intelligence, information, capital and speed. But the advent and insidious spread of the black boxes has begun squeezing humans out of their ecological niche in the financial world. These computers, backed by teams of mathematicians, engineers and physicists ('quants', they are called) and billions in capital, operate on a time scale that even an elite athlete could not comprehend. A black box can take in a wide array of price data, analyse it for anomalies or statistical patterns, and select and execute a trade, all in under 10 milliseconds. Some boxes have shaved this time down to two or three milliseconds, and the next generation will operate on the order of microseconds, millionths of a second. The speeds now dealt with in the markets are so fast that the physical location of a computer affects its success in executing a trade. A hedge fund in London, for example, trading the Chicago Board of Trade, lags at least 40 milliseconds behind the market, because that is the time it takes for a signal, travelling at close to the speed of light, to travel back and forth between the two cities while a price is communicated and a trade executed, and the delays added by routers along the way mean the actual time is considerably longer. Most companies running boxes therefore co-locate their servers to the exchange they trade, to minimise the travel time for an electronic signal.

Many of these boxes are what are called 'execution-only' boxes. This type of box does not look for trades, it merely mechanises their execution. At this task, boxes excel. They can take a large block of equities, for example, and sell it in pieces here and there, minimising the effect on prices. They test the waters, looking for deep pools of liquidity, a practice known as pinging, just like a sonar searching the depths. When they find large bids hidden just below the surface of existing prices they execute a block of the trade. In this way they can move enormous blocks of stock without rippling the market. At this trading exercise, boxes are more efficient than humans, faster and nimbler. They do what Martin did when he pieced out of the DuPont trade, only they do it better. Many managers have started to ask why traders spend so much time and effort executing client trades when a box could do it just as well, and never argue over its bonus.

Other boxes do more than execution: they think for themselves. Employing cutting-edge mathematical tools such as genetic algorithms, boxes can now learn. Funds running them regularly employ the best programmers, code-breakers, even linguists, so the boxes can parse news stories, download economic releases, interpret them and trade on them, all before a human can finish reading even a single line of text. Their success has led to an exponential growth in the capital backing them, and boxes already make up the majority of trading by volume on many of the largest stock exchanges; and they are now spreading into the currency and bond markets. Their growing dominance in the markets is one of the most significant changes ever to take place in the markets. I, like many others, find the markets increasingly inhuman, and when I trade now I often have difficulty catching the scent of the market's trail.

Human traders such as Martin are therefore in a fight for their lives. Unbeknownst to outsiders, every day a battle rages up and down Wall Street between man and machine. Some informed observers believe human traders have had their day, and will meet the same fate as John Henry, the legendary nineteenth-century railway worker who challenged a steam drill to a competition and ended up rupturing his heart.

Others, however, note with optimism that human traders are more flexible than a black box, are better at learning, especially at forming long-term views on the market, and thus in many circumstances remain faster. Evidence of their greater flexibility is found when market volatility picks up after some catastrophic event, like a credit crisis. Then managers at the banks and hedge funds are forced to unplug many of their boxes, especially those engaged in medium- and long-term price prediction, as the algorithms fail to comprehend the new data and begin to lose ever increasing amounts of money. Humans quickly step into the breach.

Something much like this occurred during the credit crisis of 2007–08. Anecdotal evidence and published fund performance statistics give us something like the following scorecard: in high-frequency trading, humans and machines fought to a draw, both making historic amounts of money; in medium-term price prediction, in other words seconds to minutes, humans pulled slightly ahead of the boxes, as flow traders made record amounts of money; but in medium- to long-term price prediction, minutes to hours or days – the boxes engaged in these time horizons are known as statistical arbitrage and quantitative equity – humans outperformed the boxes, because only they understood the implications of the political decisions being made by central bankers and Treasury officials. Thus, in what may have been the first major test of human versus machine trading, humans won, but only just. And so it is that this futuristic battle ebbs and flows.

Whatever the outcome of that battle, the financial landscape on which human risk-takers conduct their searches has been changed forever by the arrival of these machines. Governments and regulators fear the changes, suspect that the speed and opacity of the algorithms could lead to uncontrollable markets, even financial meltdown.

There is, however, another perspective from which to view the advent of these new and faster machines. They can be seen as a liberation for risk-takers. They may permit us to disaggregate the activity of trading into its component pieces and farm each of these out to the person or machine which does them best – the division of labour as applied to trading. As I say, time was when a trader had to possess good judgement, have a large appetite for risk, and be physically quick. Increasingly, though, especially at hedge funds, the roles of judgement and speed have been separated. Many portfolio managers are forbidden from executing their own trades, these being handed over to an execution desk, which then often uses execution-only boxes to place the trade. Even the appetite for risk can be stripped away from the trader and put in the hands of the floor manager. These developments mean that, increasingly, all that is required of a financial decision-maker is a good call on the market. With the help of machines, people in the financial world who have good judgement but who are risk-averse or dislike the physical aspects of trade execution could be fitted with what, in effect, amounts to a prosthetic risk-taker. Technology can lift Cassandra's curse.

Furthermore, if the physical requirements of trading were removed, perhaps the financial playing field would be levelled in such a way that it would not be the preserve of young men. On the trading floor of the future we could have a more even balance of men and women, young and old, selected for the quality of their judgement, with the grunt work of capital allocation and trade execution being done by computers. I will return to the important issue of women in finance later in the book.

A misunderstanding may arise from the picture of the future sketched above. We may be tempted by this and similar visions to believe that our bodies will come to play a less and less important role in financial risk-taking. I do not think that follows. Computers may indeed take over the job of quick execution of trades, but our bodies will continue to be crucial for success in the markets, because they provide us with perhaps the most important data informing our call on the market, and that is our gut feelings. Recent research in physiology and neuroscience has discovered that gut feelings are more than the stuff of legend, they are real physiological entities. Gut feelings emerge from a massive information-gathering exercise conducted by the body. And the body, as we will see, remains the most advanced black box ever created.

FOUR

Gut Feelings

The financial markets are replete with stories of hunches, instincts and gut feelings. These feelings consist, according to legend, of an inexplicable conviction that an investment is destined to make or lose money, a conviction often accompanied by physical symptoms. The symptoms reported by traders and investors are often quirky, like a coughing fit before the market goes down, an itchy elbow before it goes up. George Soros, founder of the hedge fund Quantum Capital, confessed that he relied a great deal on what he called animal instincts: 'When I was actively running the fund I suffered from backache. I used the onset of acute pain as a signal that there was something wrong in my portfolio.'

How exactly do these signals work? When we use a term like 'gut feelings' we imply that our brain receives information, valuable information apparently, from our body. We have seen in Chapter 2 how interoceptive pathways keep our brain constantly updated on the state of our body. The signals we considered, reporting on heart rate, blood pressure, body temperature, muscle tension and so on, served mostly homeostatic needs. Yet the notion of a gut feeling implies much more than this: it implies that gut feelings guide us in even the most complex mental tasks, like figuring out the stock market. How could information about heart rate, body temperature and the state of our immune system do that? What evidence is there that signals our brain receives from the body can help with our higher decisions? Recently there has been quite a bit of evidence. The signals flowing from the body to the brain act silently, hardly breaking the surface of consciousness, giving us a diffuse and barely perceptible sense of the body, but nonetheless act powerfully, influencing our every decision. Not only that, but without their guiding hand even the cold rationality of economic man cannot make any progress. Gut feelings are not only real; they are essential to rational choice.

The necessity of gut feelings becomes all the more apparent when decisions have to be made quickly, when we are online and in the flow, as Martin was this morning when given a minute or two to price the DuPont trade and then half an hour or so to buy the bonds he sold. In situations like these he does not have the leisure to gather all relevant data, consider all possible options, probability-weight their outcomes, and systematically work through a decision tree as an engineer might when given months, years even, to solve a problem. When pressed for a decision Martin therefore needs help drawing up a shortlist of options and their likely consequences. It is in this process that his gut feelings are brought in to streamline his thinking.

CAN WE TRUST OUR HUNCHES?

As we saw in the previous chapter, a great deal of our sensation, thinking and automatic reactions take place rapidly and pre-consciously. A number of scientists have studied the differences between pre-conscious and conscious thought, and have given these two sorts of thinking some memorable names. Daniel Kahneman calls them fast and slow thinking; Arie Kruglanski and colleagues, emphasising the motor element of thought, call them locomotion and assessment; others call them hot and cold decision-making. I prefer to think of them as online and offline thinking. Colin Camerer, George Loewenstein and Drazen Prelec, three of the founders of the new field of neuro-economics, have surveyed this research and summarised the differences between the two types of brain processing, labelling them automatic and controlled thought.

Automatic thought	Controlled thought
Involuntary	Voluntary
Effortless	Effortful
Proceeds in parallel; many steps carried out simultaneously	Proceeds serially; one step at a time
Largely opaque to introspection; we cannot trace the mental steps we followed when reaching a conclusion	Largely open to inspection; we can retrace the mental steps we followed when reaching a conclusion

Most of our thinking, they point out, takes place automatically, humming along behind the scenes, quietly, efficiently and rapidly.

A nice illustration of automatic thinking can be found in an experiment conducted by Pawel Lewicki and colleagues in which they asked people to predict the location on a computer screen of a cross that would appear at differing spots and then disappear. Unbeknownst to the participants, the location of the cross followed a rule, so it could be predicted. However, the rule was so complicated that no participant could formulate it explicitly. Yet, despite their inability to say what this rule was, people got better at predicting the location of the cross. In other words, the participants were learning the rule preconsciously. This is a lovely experiment. It demonstrates that many of the mental processes we commonly assume are conscious in fact take place below the surface of awareness.

The intuitions of traders most likely rely on just this sort of pre-conscious processing of correlations. When making a statement such as this I have to tread carefully, for buried here is a minefield of issues. To begin with, many economists and cognitive scientists have disputed the supposed reliability of intuition and gut feelings. Can we trust judgements, they ask, that simply pop into our heads? Are gut feelings really the oracular deliverances they are often claimed to be? Behavioural economists think not. They have convincingly and in great detail shown that much of our automatic thinking comes warped by biases that frequently get us into trouble. Others, most notably the German psychologist Gird Gigerenzer, respond that many of our automatic thinking patterns are in fact efficient adaptations to real-life problems. Nonetheless, the issue remains: if gut feelings are sometimes right and sometimes wrong, then how can we know when to trust them? If we cannot know, then frankly intuitions are not of much use. We should instead, argue many economists, psychologists and philosophers, use more controlled thinking, bring in the correctives of logic and statistical analysis, to overcome the shortcomings of first impressions.

In order to answer the question of whether we can trust intuitions, we should first recognise that intuition is not an occult gift – it is a skill. An insightful answer along these lines emerged from what began as a dispute and developed into a collaboration between Daniel Kahneman and Gary Klein, a psychologist who studies naturalistic decision-making, in other words decisions made out in the field by experts. At first Kahneman doubted the reliability of intuition, while Klein believed in it. As they hashed out their disagreements, it became apparent that their different views stemmed from the types of people they were studying. Klein was working with people who had developed an expertise in fast decision-making – firefighters, paramedics, fighter pilots – and who unquestionably did possess intuitions worth trusting. Kahneman, for his part, was working with people whose predictions performed no better than chance – social scientists, political forecasters, stock pickers – people we should listen to only with a healthy dose of scepticism. So what separates these two groups? Why does one develop skill and reliable intuition, while the other does not?

Kahneman and Klein first agreed that intuition is the recognition of patterns. When we develop a skill at some game or activity we build up a memory bank of patterns we have lived through, and of which we have seen the consequences. Later, when encountering a new situation, we rapidly scroll through our files looking for a stored pattern that most closely resembles the new one. Chess grandmasters, for example, are said to store up to 10,000 board configurations which they access for clues on what to do next. Intuition is thus nothing more mysterious than recognition.

Given this point, Kahneman and Klein went on to conclude that intuitions can be relied on only if two conditions are met: first, people can develop an expertise only if they work in an environment that is regular enough to produce repeating patterns; and second, they must encounter the patterns frequently and receive feedback on their performance quickly, for only in this way can they learn. Playing chess exemplifies these conditions: chess grandmasters play game after game, the rules are fixed, and they find out quickly if their moves were right or not. Much the same can be said of paramedics, firefighters and fighter pilots. Political forecasters, on the other hand, inhabit a world that is far too fluid and complex to produce patterns, and even if something like a pattern does emerge, it does so with such a long time lag that learning it may take a lifetime. 'Remember this rule,' advises Kahneman: 'Intuition cannot be trusted in the absence of stable regularities in the environment.'

The question for us becomes, do the financial markets present stable regularities? Only if they do can traders and investors rely on their hunches. Within economics, opinions on this question have been close to unanimous: the markets do not. The strongest statement to this effect comes from the Efficient Markets Hypothesis in economics. According to the economists arguing for this hypothesis, the market moves when new information arrives, and since news by its very nature cannot be predicted, neither can the market. The legend of traders and investors heroically drawing on gut feelings is, they argue, pure mythology. No one can predict the market, nor consistently outperform it.

But is this true? The Yale economist Robert Shiller suspects it is not. He does not buy the idea that nothing – no personal trait, no training – can improve a trader's performance. Shiller believes, on the contrary, that investing is like any other occupation, and that intelligence, education, training and hard work can indeed improve your performance. I think he is right. I also suspect that the Efficient Market Hypothesis has been a boon to many of the physicists, engineers and code-breakers employed by hedge funds, for these scientists have been able to find tradable patterns in what efficient market theorists believed was pure noise, and to build algorithms to exploit these patterns. Efficient market theory, because it has been orthodoxy for decades, may have limited the number of competitors looking for these patterns.

My experience with traders has been that they can indeed learn patterns, that they can develop an expertise in predicting the market. I and a colleague, Lionel Page, a brilliant statistician and behavioural economist, tested this hypothesis by looking at how consistently a group of traders made money, consistency being determined by what is known in finance as their Sharpe Ratio. The idea behind this measure is simple: it measures how much risk was taken in making a given amount of money. For example, if one trader makes $100 million in a year, and in the course of doing so never makes or loses more than $5 million in a single day, then his performance has been steady, his risk low, and his Sharpe Ratio would be high. If another trader makes $100 million but alternates between making $500 million one day and losing it the next, then his profit looks like no more than luck, even dumb luck, his risk is far too high, and his Sharpe Ratio would be low.

The differences between these traders can be compared to the driving styles of two taxi drivers. The first holds to the speed limit and gets you to the airport in 45 minutes. The second speeds at 100 mph for 15 minutes, stops for a coffee, drives back 10 miles to pick up a newspaper, then floors it at 120 mph into oncoming cars, snarls traffic for hours afterwards and narrowly avoids several head-on collisions, but by some miracle arrives at the airport in 45 minutes. 'See, I told you I'd get you there on time,' he says, and with that he holds out his hand for a bonus – I mean tip. Now, which of these drivers are you going to tip? Or ride with again? Sharpe Ratios allow the banks in effect to give their traders a driving test.

According to the efficient market hypothesis, traders and investors cannot make money more consistently than the stock market itself. This statement is comparable to saying you cannot drive to the airport – obeying the speed limit, that is – in under 45 minutes. But what we found in our study was that the traders could. They were like crafty cab drivers who keep figuring out shorter routes to the airport. The

S&P 500 (an index of the prices of 500 large US stocks) has a long-term Sharpe Ratio of about 0.4, yet the experienced traders in our study had Sharpes higher than 1.0, the gold standard among hedge funds.

Were they lucky, or skilled? The question has more than an academic interest. Banks and hedge funds have to decide how to allocate capital, risk limits and bonuses among their traders, so it is crucial that they be able to distinguish luck from skill. During the credit crisis of 2007–09 bank managers found out, to their and everyone else's dismay, that most of their star traders turned out to be like the crazy cab driver, and they lost more money in these two years than they had made in the previous five. Could the banks not have distinguished luck from skill?

Our data indicated that they could. We found that the experienced traders who consistently made money, even through the credit crisis, were ones whose Sharpe Ratios had risen over their careers. When we plotted their ratios against the number of years they had been in the business we found a nice upward slope, indicating that they had been learning to make more money with less risk. Shiller was right: training and effort do pay off in the markets. This finding led us to suggest that banks and hedge funds could determine which traders had developed a skill worth paying for merely by looking at the slope of their Sharpe Ratios plotted against the number of years they had traded. If the curve was upward-sloping, chances are this trader had developed an expertise in pattern recognition that was indeed worth paying for.

A couple of points follow from this discussion of intuition and hunches and our data on the Sharpe Ratio. First, there does appear to be such a thing as trading skill. The financial markets seem to meet Kahneman and Klein's criteria for an environment in which intuitions can be trusted. It is not surprising therefore that discussions between traders frequently compare the current market to something they have traded through before – 'This credit crisis feels just like the Russian default in '98. I bet the yen rallies.' Most of a trader's dialogue, however, remains internal and pre-conscious, a good trader listening attentively to whispers from the past.

Second, the question that often crops up in discussions of intuition – which is more reliable, intuition or conscious rationality? – is something of a red herring. We inevitably use both. If we tried to use nothing but conscious rationality, tried to emulate Spock, we would find ourselves uncontrollably thrown back for most of our decisions on fast, online thinking. The question of whether the resulting decisions are trustworthy is not one that can be answered once and for all – 'Yes, always trust your hunches,' or 'No, work out a decision tree.' Its answer will depend on your training. We should not be asking if we should trust our intuitions; we should be asking how we can train ourselves to possess a skill that can be relied on.

HUNCHES AND GUT FEELINGS

But what, you may ask, do intuitions and pre-conscious processing have to do with gut feelings? Just because a mental process takes place behind conscious awareness does not mean it is drawing on signals from the body. Indeed, pattern recognition, even though silent, probably draws on higher brain regions, parts of the neo-cortex and the hippocampus, a brain region acting as the filing system for memories. What is the connection between these higher brain regions and the body? There is in fact a connection between pre-conscious decisions and the body, because it is gut feelings that allow us to rapidly assess whether a pattern and a considered choice will most likely lead to a pleasant or a nasty outcome, whether we like or dislike, welcome or fear it. Without such visceral colouring we would be lost in a sea of possibilities, unable to choose – a situation the cognitive psychologist Dylan Evans has called 'the Hamlet Problem'. We may be gifted with considerable rational powers, but to solve a problem with them we must first be able to narrow down the potentially limitless amount of information, options and consequences. We face a tricky problem of limiting our search, and to solve it we rely on emotions and gut feelings.

Such is the conclusion drawn by two neuroscientists, Antonio Damasio and Antoine Bechara. Working with patients who had damage to one particular part of the brain which integrates signals from the body, Damasio and Bechara found that these patients could have perfectly normal, even exceptional, cognitive abilities, yet make terrible decisions in their lives. Perhaps, Damasio and Bechara surmised, the patients' IQs counted for little in making good decisions because they were deprived of help from their bodies, from homeostatic and emotional feedback. They concluded that 'feeling was an integral component of the machinery of reason'.

To account for the decision-making impairments of these and similar patients, Damasio and Bechara developed their 'Somatic Marker Hypothesis'. According to this hypothesis, each event we store in memory comes bookmarked with the bodily sensations – Damasio and Bechara call these 'somatic markers' – we felt at the time of living through it for the first time; and these help us decide what to do when we find ourselves in a similar situation. When we scroll through the options open to us, each may present itself with a subtle tensing of muscles, a quickening of breath, a slight shiver of dread, a brief moment of calm, a frisson of excitement – until one just feels right. These bookmarks are especially memorable, and most urgently needed, when we take a risk, for risk can hurt us, physically and financially. So it is little wonder that tales of gut feelings are so frequent and legendary in the financial markets.

Scientific research into gut feelings represents a new perspective on the brain and body. Scientists such as Damasio and Bechara argue that rationality by itself gains no purchase on the world, it merely spins its wheels, without the grit of somatic markers. They draw our attention to the physical aspect of thought, and raise the possibility that good judgement may require the ability to listen carefully to feedback from the body. Some people may be better at this than others, may have more efficient connective circuitry between body and brain, just as some people can run faster than others. On any Wall Street trading floor you will find high-IQ, Ivy League-educated stars who cannot make any money at all, for all their convincing analyses; while across the aisle sits a trader with an undistinguished degree from an unknown university, who cannot keep up with the latest analytics, but who consistently prints money, to the bafflement and irritation of his seemingly more gifted colleagues. It is possible, though odd to contemplate, that the better judgement of the money-making trader may owe something to his or her ability to produce bodily signals, and equally to listen to them. We tend to think – we want to think – that decisions are a matter of cognition, of mind alone, pure reason, a view Damasio calls 'Descartes' Error'. But good judgement may be a trait as physical as kicking a football.

An interesting possibility arises: could we tell whether one person has better gut feelings than another? Could we monitor feedback from their bodies? Gut feelings, like the oracle at Delphi, provide valuable insights, but are frustratingly hard to access and notoriously hard to interpret. This inaccessibility is due in part to their being processed by regions of the brain that are not fully open to conscious inspection. Could we access these signals in some way other than introspection? Could we some day hack into these communication lines between body and brain, and then use this information as a trading signal?

FEEDBACK

We all recognise that our thoughts affect our bodies. To take the most trivial example, it is your brain that tells your hand to reach for a glass of water sitting on the kitchen table. But so too does your body affect your thoughts, and here again everyday examples are easy to find. When you are hungry or thirsty, for instance, your thoughts change and you develop what is called a 'selective attention' to signs of food and water, and you stop paying attention to anything else, such as the book you are reading or the beauty of a sunset. Other examples of how the body affects the brain are less familiar, yet, if you stop to consider them, should be equally obvious. Like the fact that your brain, like a muscle, requires blood, glucose and oxygen to operate. In fact your brain, which constitutes only 2 per cent of your body mass, consumes some 20 per cent of your daily energy. You can verify the sobering fact that thinking is a physical process by monitoring the pulse of your carotid artery, which supplies blood to the brain. Engage in a taxing mental task, such as mental arithmetic, and as you do so press two fingers gently into your neck just below the corner of your jaw: you can actually feel your pulse speed up as the machine in your head draws more fuel.

In a more formal experiment, a group of radiologists in Miami measured glucose use by the brain during a verbal fluency task, which required participants to list as many words as possible beginning with a given letter in a short space of time. They found that people performing this simple task drew 23 per cent more glucose into their brains than they did when at rest.

Disturbingly, a group of psychologists at Florida State, also looking at glucose levels in the brain, found that during taxing mental (as well as physical) activities our glucose reserves become depleted, and this reduces our capacity for self-control. They concluded that allocation of energetic resources during emergencies follows a 'last in, first out' rule, according to which mental abilities that developed last in our evolutionary history, like self-control, are the first to be rationed when fuel is low. Muscles, which draw a small amount of glucose when at rest, come to monopolise available resources during physical activity. The preferential treatment of muscles during a fight, or when playing a sport, for that matter, and the rationing of glucose to the brain regions responsible for self-control, might explain why fights so easily run out of control (ice hockey seems especially prone). Perhaps the same could be said of our self-control when we are working long hours at the office – we tend to snap more easily – or trying to stick to a diet, since the draining of glucose also drains us of resolve.

As I say, these simple facts about how the brain affects the body and the body the brain are ones we all recognise. But things get a bit stranger when we consider one of the most poorly understood and under-researched phenomena in all the neurosciences – feedback between body and brain. In the examples we have just looked at, the brain affects the body, or the other way round, but the flow of causation is one-way:

brain affects body, for example, and that is the end of the story. But the situation is very different in cases of body–brain feedback. In feedback a thought affects the body, and the changes taking place in the body then feed back on the brain, changing the way it thinks.

Take a simple example of the process: when you feel depressed you may decide, in a moment of self-assertion, to pick yourself up and battle on, forcing a smile, straightening your posture, walking more briskly; and in time these changes may actually work, you may end up feeling happier. Here changes in your body – its posture, facial expression, etc. – have fed back on your brain and changed the thoughts you think. Body–brain feedback of this sort can even, under some circumstances, turn into runaway reactions which end in extreme behaviour. When you are scared, for example, your heart beats faster, you sweat, hyperventilate, and run away from a supposed threat. As you become aware of these symptoms you can start to worry even more, and the stage is set for a fully-fledged panic attack. Or an argument, to take another example, may escalate into pushing, and in the extreme to an exchange of blows. At each stage blood pressure increases, breathing accelerates and, importantly, brains lose their cool. Even a gentle person, as the physical exchange progresses, can begin to think violent thoughts. This spiralling into out-of-control thoughts and behaviour can also work in a more welcome direction, as it does for example during sex. In all these examples of feedback, the brain does not look on as a disinterested observer of a body in turmoil, it is intimately caught up in the process. It is participant, not spectator.

Nowhere has this feedback been more accurately and insightfully described than in the writings of William James, the great nineteenth-century philosopher and psychologist (and brother of the novelist Henry James). 'Everyone knows how panic is increased by flight,' he wrote, 'and how the giving way to the symptoms of grief or anger increases those passions themselves. Each fit of sobbing makes the sorrow more acute, and calls forth another fit stronger still, until at last repose only ensues with lassitude and with the apparent exhaustion of the machinery. In rage, it is notorious how we "work ourselves up" to a climax by repeated outbreaks of expression.'

Words we all recognise as true; yet the familiarity of the experience James described hides a mystery: why are we built this way? If your brain wants to cheer itself up, or worry itself into a panic, why bother sending its message through the body? Why not send a signal directly from one brain region to another? These questions, in my view, take us right to the heart of the mind–body problem. For the body does indeed influence the brain, transforms its very thoughts and feelings. But again, why?

If the brain kept its signalling within the brain, surely it would save a great deal of time. Maybe. But probably not. Not if the point of the thought is to produce movement, for then the extra processing time in the brain might actually slow down our eventual action.

This was largely the conclusion drawn by James. He came to it when puzzling over a very special and powerful form of gut feeling – our emotions. He had begun to suspect that we commonly misunderstand the nature of our emotions: we tend to think that our emotional feelings come first, and then cause our emotional behaviour. But according to James, the feeling of an emotion is in some ways the least important part of the experience. In this our common-sense understanding of emotion has it all wrong. We tend to think that we cry because we are sad, run from a bear because we are scared, but James argued that the true course of events runs the other way round. We are sad because we are crying, scared because we are running away. More precisely, the course of events unfolds as follows: we perceive a bear; our brain triggers automatic escape behaviour, such as running; and this physiological change is then reported back to our brain and shows up in our consciousness as a feeling of fear. Some scientists have even argued that the feeling itself plays a very small role in an emotional event, tagging along as mere observer of action already taken. It is, as the neuroscientist Joe LeDoux suggests, no more than 'icing on the cake'.

James was trying to correct the belief prevailing at the time that an emotion is largely a mental event, like a thought, only with strong feelings attached. Arguing against this view, he pointed out that it missed what was most important about emotions: that they are first and foremost reflexes designed to help us behave and move, sometimes quickly, at crucial moments of our lives. As Charles Sherrington, the Nobel laureate in physiology for 1932, was later to say, 'emotions move us', and he meant literally. If the functional role of emotion is to promote fast behavioural reactions, then what are the emotional feelings? James thought that they emerge when we perceive the changes our body goes through during an emotional episode, changes such as a tense stomach, sweating, a pounding heart, increased blood pressure, raised body temperature and so on. In the absence of these physical sensations, emotions would be drained of feeling. 'Without the bodily states following on the perception,' James wrote, 'the latter would be purely cognitive in form, pale, colourless, destitute of emotional warmth. We might then see the bear, and judge it best to run, receive the insult and deem it right to strike, but we could not actually *feel* afraid or angry.'

Casual support for James's view can be found in our everyday use of emotional language, for we draw heavily on bodily metaphors: we speak, for example, of receiving bad news with a sinking stomach, a broken heart, or blood-boiling anger; we tell of a chilling moment, a tense encounter, a heart-pounding experience, of being hot under the collar or flushed with excitement.

James's account of the stages of emotional feedback – physical reaction first, conscious feeling second – may seem counter-intuitive. But it makes perfect sense if what we need in an emotional crisis are fast reactions: we act first, feel later. Still, James received a lot of criticism for his theory, most notably during the 1920s from the great Harvard physiologist Walter Cannon. James versus Cannon was a clash of titans. Cannon argued that feedback from the body travelled far too slowly to keep up with the rapidly changing feelings you can have in the course of an emotional encounter, one that may see you pass from anger to fear to relief to happiness, all within the space of a few seconds. For the suite of bodily changes, including fluctuations in breathing, body temperature and adrenalin levels, to keep pace with each fleeting nuance of feelings, they would have to increase and decrease with split-second timing. But they do not. Some of these physiological changes, such as adrenalin release, can take one or two seconds before they are felt, so our visceral organs would be left behind by high-speed emotional events.

There was another problem with James's theory, or so Cannon argued. Cannon thought that feedback from the body was not unique enough to provide an individual physiological signature for each emotion. Your heart pounds and your breathing accelerates when you are scared, or angry, or joyous, or when you fall in love. In fact, whenever you experience a strong emotion, argued Cannon, you produce much the same suite of physical reactions. He reported, for instance, 'the case of a young man who on hearing that a fortune had just been left him, became pale, then exhilarated, and after various expressions of joyous feeling vomited the half-digested contents of his stomach'. He tells of similar symptoms of nervous agitation displayed by people suffering deep sorrow or great disgust. Bodily feedback may increase your arousal during an emotional encounter, but it cannot tell you which emotion you are feeling. Physical arousal, Cannon concluded, is too ham-fisted to paint the often gentle hues colouring our emotional life.

In the end Cannon's arguments carried the day, and James's theory retreated from the field of emotional studies, living on in the nether world of interesting but disproved ideas. Yet in the 1970s and 80s that began to change. Many scientists took a renewed interest in feedback between body and brain, and decided it was time to take another look at James's theory.

What they found was that Cannon had formulated his criticisms by focusing exclusively on what can be called the visceral nervous system, the network of nerve fibres controlling your heart, lungs, arteries, gut, bladder, sweat glands and so on. But the visceral nervous system is just one of many lines of communication operating between body and brain. Indeed, it is not even the whole of the nervous system, for in addition to the nerves connecting brain to visceral organs there is the nervous system connecting brain to skeletal muscle, and this system employs signals that move at lightning speeds. Recent research has found that signals from the body do travel fast enough to generate our high-speed emotional life, and are complex enough to generate its richness. Let us look at these two points in turn.

MUSCLES AND OUR FIRST RESPONSE

When the body wants to send a signal at high speed it uses electrical signals rather than blood-borne chemical ones like hormones. But nerve fibres vary dramatically in their speed of transmission, so the body and brain choose carefully the fibres they entrust with a message. The fibres of the nervous system that connect visceral organs to brain are relatively slow, carrying their signals at speeds ranging from 5 to 30 metres per second, with some ambling along at a mere one metre per second. However, the muscular nervous system is made up of a different class of fibre altogether, and these carry signals at close to 120 metres per second. If we were to compare our body's wiring to the internet, then the visceral nervous system constitutes a 56k modem and the muscular nervous system its broadband, the closest thing we have to instant messaging. This feature of our bodies makes perfect sense, for it is the speed of movement during emergencies that keeps us alive.

It also turns out that our muscles play an intimate role in our emotional expressions. When we are angry or sad or elated our posture changes, and muscles in one part of our body tense while others relax. The muscular nervous system, moreover, is fast enough to keep up with, even to cause, our fluctuating emotional feelings.

One set of muscles in particular has been found to play a central role in our emotional lives – the facial muscles. Some of the most exciting work on emotions and body–brain feedback has involved studying facial expressions, in particular a class called micro-expressions. These were discovered during the 1960s by William Condon at the University of Pittsburgh, among others, when he studied slow-motion film of patients undergoing psychotherapy. Condon was astonished to find facial displays of anger, disgust, fear and other emotions flicker into life and then vanish, all within the space of 40 milliseconds – that is, a mere twenty-fifth of a second. These expressions come and go so quickly we are not even aware that we have made them. But they carry a load of meaning. Their study was later taken up by Paul Ekman, a psychologist at the University of California, who began training police and security services to spot these micro-expressions as a new and reliable method of lie detection.

Faces are objects of unique significance to us, and to many other mammals as well. It is largely through faces that we learn of other people's intentions, and they ours. When we are angry we broadcast our threat, and when sad our need for reassurance. When we encounter a person we usually begin by examining their face, either directly or surreptitiously, while they do the same to us. The result is a silent exchange in which we discern if this person is friend or foe, if we trust or distrust them; and after a moment or two the exchange may settle into a stable interpretation on both sides – we like each other. We are often only dimly aware of the changing weather on our faces as we shift from interpretation to interpretation of the person in front of us. But we also try to fool people by disguising our true feelings with the mask of another emotion, as do people looking back at us. A salesman may smile winningly at us but feel nothing but contempt.

Micro-expressions play a key role in maintaining a line of truth in this game of facial spy and counter-spy. The salesman's micro-expression may betray his duplicity. We have little control over micro-expressions, so in many ways they remain a true gauge of our real feelings and intentions. Since mistaking foe for friend can be fatal, our brains have been built to process information coming from faces faster than from pretty well any other object in the world. Micro-expressions break the surface of our faces, transmit their signal, and then submerge just as quickly, all within 40 milliseconds; but an observer can register these signals in as little as 30 milliseconds, far faster than their conscious awareness. These extraordinary speeds mean we could potentially have an entire conversation, with several rounds of micro-expression and response, within the space of a single second, and all without any awareness it has taken place. We may merely walk away from a brief encounter with a stranger nagged by a vague uneasiness.

The speed of our muscular reactions in general, and the almost incredible speeds of facial expressions in particular, have led many researchers to venture what has been called the 'facial feedback theory of emotions', according to which the purpose of facial expressions is not so much to express feelings, as to generate them. This new theory echoes that of James: we act first, feel later. If this theory is true – and a great deal of research now suggests it is – it raises a number of intriguing questions. For example, do people with very expressive faces – people who have been delightfully called 'facial athletes' – experience a richer emotional life? Are the tight-lipped Brits emotionally handicapped? Hard to say. It is possible that people with more labile faces simply become habituated to their facial antics, and that a poker-faced Brit might succumb to an outwardly unseen emotional torrent caused by little more than a twitch of the mouth or a furtive glance. On the other hand, people who inject botox into their cheeks, foreheads and eye creases, thereby anaesthetising their facial muscles, may be dampening their emotional and indeed their cognitive reactions. Ironically, it is often movie actors who do so, yet if there is any truth to the theory of Method acting, according to which you should conjure up a real emotion rather than artfully fake it, then these actors may be killing their very talent.

Robert Levenson and Paul Ekman, two psychologists working on emotional display, have conducted a series of fun experiments to demonstrate how feedback from facial expressions alone can bring about a range of emotional feelings. They instructed participants to flex this muscle or that in their faces, relax another muscle, or to hold a pencil at the back of their teeth. While following these instructions the participants would, without knowing it, compose an emotional face, say one of happiness or sadness. After this purely physical exercise they were tested for mood. Levenson and Ekman found that by moving their facial muscles alone, without any emotive input, the subjects had come to feel the mood portrayed on their faces.

Extraordinary research. In fact, just as William James had predicted. He too recognised that muscles can communicate an emotional feeling to the brain. Even when our muscles appear outwardly unchanged, he wrote, 'their inward tension alters to suit each varying mood, and is felt as a difference of tone or of strain. In depression the flexors tend to prevail; in elation or belligerent excitement the extensors take the lead.'

YOUR GUT IS TELLING YOU SOMETHING

Our body, through its muscles, can thus transmit information back to the brain fast enough not only to keep up with our emotional life but also to generate it. Furthermore, our body can compose messages that are complex enough to produce the full range of our emotions. It does so by drawing on a wide palette of signals, electrical ones sent by muscles and by our visceral organs, and hormonal ones carried by the blood. Contrary to Cannon's view, our body has so many different signals at its disposal that together they can easily compose messages with all the subtlety of a piano keyboard, and some with the speed of a radio transmission.

The various electrical and chemical systems carrying these signals are brought online in sequential order as a challenging event unfolds. Our muscles, especially our facial muscles, kick in quickly and unreflectively, in a matter of milliseconds. Shortly thereafter the visceral nervous system, operating on the order of milliseconds to seconds, calls into action the tissues and organs, such as lungs, liver, adrenal glands, that will support our muscles during the crisis. Moments after these two electrical systems have been brought online our chemical systems begin to switch on. Fast-acting hormones like adrenalin, released in seconds to minutes, flood into the blood and unpack energy stores for immediate use. Finally, if a challenge persists, then our steroid hormones take charge, and over the course of hours, even days, they prepare our bodies for a change of life. At this point our bodies retool, girding for attack or hunkering down for a siege. Each of these staggered physical changes is reported back to the brain, where it alters our emotions, moods, memories and thoughts. To see in a highly simplified way how these feedback loops work, let us watch Gwen, the trader sitting next to Martin, deal with a scare.

Gwen, a former college tennis star with a brief stint on the professional circuit (her best year took her to the last 16 at the Australian Open), now trades five-year Treasuries. She has a solid track record at making money, but for the past month or so she has been in a slump. Normally that is no big deal – all traders go through periods when they do not make any money. Nonetheless, no trader feels comfortable at these times. There is a saying on Wall Street: you're only as good as your last trade. Well, shortly after the DuPont trade Gwen follows Martin down the aisle to the coffee room, and on the way she catches a glimpse of Ash, the floor manager, staring at her. His facial expression broadcasts a complex message. It is almost a dispassionate gaze, but there is no denying a hint of hostility in it; and there is more, a trace of pity (why does he feel sorry for her?) and maybe disgust (the sort people feel, probably as a rationalisation, once they have decided to fire you). Gwen registers the look in a matter of milliseconds, and automatically responds with a micro-expression of shock, alarm. Muscles throughout her body tighten, straightening her posture, craning her neck. In a threatening situation such as this one, Gwen's muscular nervous system reacts first, setting off warning bells, preparing her for quick action.

As she becomes aware of Ash's glare, and her own tensed body, another set of messages starts to arrive, these from her visceral nervous system. Operating on the order of milliseconds to seconds, the visceral nervous system calls into action the tissues and organs that will support Gwen's muscles during this crisis – if there is one – providing them with fuel, oxygen, cooling, exhaust removal and so on; and, with a slight delay, it floods her arteries with adrenalin. This is the fabled fight-or-flight response, a bodywide preparation for a physical emergency, involving increased breathing, heart rate and sweating, dilated pupils, suppressed digestion, and so on. The fight-or-flight nervous system first prepares Gwen's body for action and then, by means of nerves in the spinal cord, reports her state of arousal to the brain. This information slants her perception of the world. She sees Ash's face, registers its look, and the disturbing signals from her body suggest that something is not right. *Why is he looking at me like that?*

Another part of her visceral nervous system, what is called the 'rest-and-digest' system, brings in equally valuable information, especially from her gut, perhaps gut feelings themselves. Our visceral nervous system is composed of two branches: the fight-or-flight system and the rest-and-digest system. The fight-or-flight system is brought online in times of emergency, but once the emergency passes our body needs to settle down, rest, and basically get life back to normal. It is at these times that the rest-and-digest system takes over, damping down arousal in our bodies. The fight-or-flight nerves thus work largely (but not always, as we will see in a later chapter) in opposition to those of the rest-and-digest system, the two nervous systems alternating their activities, one speeding us up, the other slowing us down. Importantly, though, both carry information back to the brain and affect our thoughts, emotions, moods.

The main nerve in the rest-and-digest nervous system is the vagus, a large and powerful nerve that exerts a calming influence on the many tissues and organs it touches. The word 'vagus' (pronounced like Vegas) is Greek for wanderer, and wander this nerve does. It emerges from the brain stem and heads down into the abdomen. In the course of its long travels it visits the voicebox, then the heart, lungs, liver and pancreas, finally terminating in the gut (fig. 6). Because of its extensive connections, this curious nerve can modulate our tone of voice, slow our breathing and heart rate, and in the stomach control the early stages of digestion. What is more, the region of the brain stem where the vagus originates is also the one that regulates our facial muscles, and this allows our facial expressions to synchronise with our heart rate and the state of our gut. By linking facial expression, voice, lungs, heart and stomach, the vagus plays a central role in our emotional lives.

It also brings messages back to the brain: almost 80 per cent of the vagus nerve's fibres (the vagus is a cable composed of thousands of

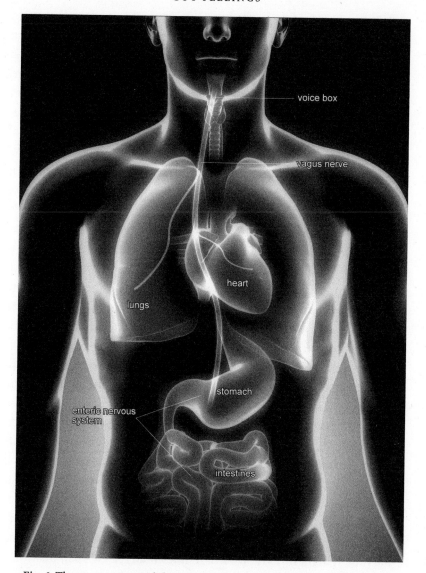

Fig. 6. The vagus nerve and the enteric nervous system. The vagus nerve, the main nerve in the rest-and-digest nervous system, links the brain stem, voicebox, lungs, heart, pancreas and gut. Eighty per cent of its fibres carry information back to the brain, mostly from the heart and gut. The enteric nervous system, often called the second brain, is an independent nervous system controlling digestion. The brain in the gut and the brain in the head communicate and cooperate (and occasionally disagree) largely by means of the vagus nerve.

fibres) carry information from body to brain. Most of this returning information comes from the gut, so one may naturally ask, do gut feelings really come from the gut? The quick answer is yes, or least some of them do. Not all, though. Interoceptive information streams into the brain from every tissue in the body, not just the gut. Nonetheless, the gut holds a special place in our physiology because, remarkably, it has its own 'brain'.

The gut is under the command of what is called the enteric nervous system (fig. 6), which controls the movement and digestion of nutrients as they pass through the stomach and intestines. Unlike other nerves in the body, this nervous system can act independently of the brain, and is one of the only systems that will continue to function even if all connection to the brain is severed. It contains approximately 100 million neurons, more than are found in the spinal cord, and produces the same neurotransmitters as the brain. The enteric nervous system has been aptly termed by Michael Gershon 'the Second Brain'. And it is the vagus nerve that links our two brains, acting much like a hotline between two superpowers.

Through its control of digestive acids and enzymes, the enteric nervous system decomposes food until its constituent molecules can be absorbed into the body. I say 'into the body' because the digestive system, technically speaking, is not inside the body. The cavity inside the mouth, the oesophagus, stomach, intestines and colon remain on the outside of our body, constituting, in the words of Gershon, 'a tunnel that permits the exterior to run right through us'. The gut also powers the caterpillar-like undulations in the intestinal tube that inch food and waste forward, or rather backward. In fact it was the discovery of these undulations that led to the further discovery of the enteric nervous system. In 1917, Ulrich Trendelenburg, a German physiologist, removed a section of intestines from a guinea pig, severing all connection to the brain. When he blew into this section he was amazed to find the air blowing right back. This was not the sort of blowback you would get if you blew into a balloon and it squeezed the air back out. This was different. After a moment's delay the intestinal section contracted and puffed a light gust of air right back at

Trendelenburg, like some gentle creature playing a simple game. At that point it dawned on Trendelenburg that what he was dealing with was an independent nervous system.

The brain and the enteric nervous system, being connected by the vagus, send messages back and forth, affecting their respective decisions. Conditions in one brain may show up as symptoms in the other. For instance, when stressed, the brain in our head may inform the brain in our gut of an impending threat and advise it to stop digesting, such digestion representing a needless drain on energy. To take further examples, patients with Alzheimer's often suffer constipation, as do people addicted to opiates; while patients on antidepressants often experience diarrhoea. Information may also flow the other way, with events in the gut causing changes in the brain. For example, people suffering from Crohn's disease, a form of inflammatory bowel disease, are more easily aroused by emotional stimuli. Furthermore, hormones secreted in the gut during feeding can enhance the formation of memories, the evolutionary rationale being, I suppose, that if you have eaten some food, then your gut hormones instruct your brain to remember where you found it. Of course, the effects of eating can also be highly soothing: a good meal can prove more than a mere gustatory treat: it can settle the body and calm the brain and suffuse us with a profound sense of well-being. In short, neural activity in our head can affect our digestion; neural activity in our gut can affect our mood and thoughts.

Gwen feels her stomach knot, her breathing speed up, her heart pound a bit harder, and these feelings, funnelled into the brain by the vagus, slant her interpretation of Ash's glowering look. She accordingly experiences a moment of fear. But fortunately, not for long. Ash breaks off his look and turns away. Gwen thinks through the encounter and tells herself not to be so silly – he was probably looking straight through her, thinking of something else, maybe a bad position on the mortgage desk, maybe his all-too-public marital problems. She shakes off her worries, her body begins to settle, and she continues to the coffee room, not giving the incident another thought.

But pre-conscious parts of her brain and her body are not quite so convinced. Pre-consciously, other information is being weighed: rumours of a reorganisation of the desks, a joke Ash made at her expense at a recent client dinner. Fifteen minutes later, coffee in hand, she recalls Ash's look and her stomach knots once again. This time she cannot shake off her concerns. Things are starting to add up; she suspects she is going to be reassigned. But where? Why?

Gwen now faces a long-term challenge, and to deal with it her steroid hormones take charge. That is what steroids do: they prepare the body for a change of behaviour. For example, should she encounter a situation of extraordinary opportunity, such as a bull market, then testosterone, produced by both ovaries and adrenal glands, takes charge and prepares her body for an extended period of competition. If, however, she finds herself faced by an uncontrollable stressor, such as a market crash or an angry boss, then cortisol organises a coherent long-term physical defence. Steroids, acting over the course of hours, even days, are the final, slowest and most comprehensive step in our body's graded response to a challenge. Gwen may be blessed with an admirably toughened physiology, and her extensive experience on the tennis circuit makes her almost unshakeable in the face of risk. But not office politics. Office politics unnerve her. She hates them. Over the next few hours – days, if this thing is not resolved soon – under the influence of ever higher levels of cortisol, she develops a mood called anticipatory angst, and it vexes her every waking minute.

A number of points emerge from this scenario. To begin with, Gwen's bodily feedback during this emotional encounter is not confined to her fight-or-flight nervous system, as Cannon argued it was. Messages from her body are carried by muscles, fight-or-flight nervous system, rest-and-digest nervous system, and hormones, and are in fact diverse and subtle enough to transmit a rich emotional life (see fig. 7). In fact, many scientists have found that each of our emotions is tagged with a distinct pattern of nervous and hormonal activation. Gwen tailors her heart rate, muscle tension, digestion, vascular resistance, sweating, bronchial contraction, blushing, pupil dilation, facial expression and so on to each situation.

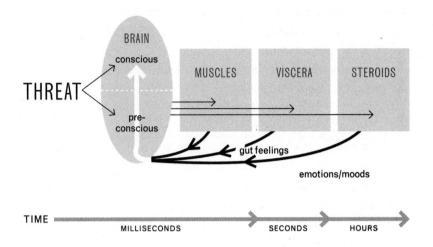

Fig. 7. Gut feelings and feedback loops between body and brain.

These physiological reactions then feed back on Gwen's brain. But what her brain experiences is not pure observation; she does not watch her body in a disinterested way. She experiences this feedback in the form of an emotion or a mood. Emotions and moods are different; they operate over different time scales. Emotions are short-lived. It has been suggested that emotions are designed to be fleeting, because they provide our brain with valuable and timely information. If they were to persist, they would interfere with other, newer information being brought to our attention. A mood is slower, more like a long-term attitude, a background and slow-burning emotion which slants our view on the world. Emotions and moods both alter Gwen's attitude to events, tinker with the memories she recalls, change the way she thinks.

Pre-conscious regions of brain rapidly register threat

Muscles in body and face prepare for fight-or-flight

Visceral organs support muscles

Glands produce hormones for longer-term support of muscles

Muscle tension, heart rate, breathing, hormones, etc. send signals to brain
Pre-conscious regions of brain experience this feedback as gut feelings
Conscious regions of brain experience this feedback as short-term emotion or long-term mood
Emotion and mood ensure conscious thoughts synchronise with body to produce coherent behaviour of anger, fear, happiness, etc.

This anecdote brings us back to the question we are trying to answer: why are we built with these feedback loops? What purpose do these emotions and moods serve? Are the feelings they carry largely superfluous? Not likely. What is more likely is that these feelings help slant our attention, memory and cognitive operations so that they synchronise with our bodies. When we face an attack, for example, we want our body to be tensed and ready, but we also want our brain to think aggressively. When we want to start a family, on the other hand, we want body and brain to be in sync in a gentler, more loving manner. During important moments in our lives like these we do not want our tissues multi-tasking; we do not want, say, a body gearing up for battle but a mind thinking amorous thoughts. Feedback ensures that our tissues do not work at cross purposes. Feedback, carried by nervous system and hormones alike, unifies body and brain at the most important points in our lives. And at these moments – of euphoria, of flow, of love, of fear, of fight – body and brain merge.

THE FAINTEST FEEDBACK

Just as emotion and mood slant our thinking to suit the situation at hand, so too do gut feelings, the most subtle of bodily feedbacks. Gut feelings economise on limited computational resources and safeguard our decisions, pre-consciously steering us away from dangerous options we might be considering.

Damasio and Bechara tested for the effects of gut feelings, or what they call somatic markers, with a computer game known as the Iowa

Gambling Task. Players were presented with four decks of cards. Each card when turned over showed an amount of money the player had either made or lost. The decks had been stacked in the following way: two of the decks displayed low amounts of money, like making or losing $50 or $100, but choosing from them would, over time, lead the player to make a profit; the other two decks displayed higher amounts, say $500 or $1,000, and were therefore more exciting, but choosing from them would over time lead the player to lose money. At the beginning of the game the players did not know the properties of the decks, or even that they differed; they had to play the game naïvely and figure out how to make money as best they could.

In time, the players figured out how the decks were stacked, and which ones they should play from if they wanted to make money. But the course their learning took provided some interesting results. What Damasio and Bechara found was that players began to choose from the money-making decks before they knew why. Just as in the Lewicki experiments, where subjects had to predict the position of a cross on a computer screen, the players were learning the rule pre-consciously well before they could consciously state it. More intriguingly, their learning was guided by a signal from their bodies. While they played the game, all participants were monitored for a somatic marker, the electrical conductivity of their skin. Your skin experiences rapid and unnoticed changes in electrical conductivity, the result of momentary changes in the amount of sweat lying in its crevices. Skin conductance is highly sensitive to novelty, uncertainty and stress. The players' skin conductance began to spike when they contemplated playing from the money-losing decks, and this somatic prod proved enough to steer them away from these dangerous choices. Aided by these brief shocks, normal players were guided towards the money-making decks long before their conscious rationality had figured out why they should be doing so.

FEELING THE MARKET

As we undergo an apprenticeship in an activity like trading, we store more than patterns: we store patterns twinned with muscular and visceral reactions. When Martin is in the flow, when he encounters a market event like the DuPont trade, he has little time to carefully weigh all possible outcomes of his actions, but must volley sales enquiries quickly and profitably, and react to prices that appear fleetingly on the screens. He rapidly scrolls through patterns stored in memory looking for a match (although a perfect match is rare), and with each one his body and brain shift kaleidoscopically from one state to another. Body and brain rev up and down together. In fact, to speed up decision-making, his brain, according to Damasio and Bechara, uses predictive models, called 'as-if loops', which allow him to rapidly simulate the bodily reaction most likely to follow a considered choice of action. Relying on this as-if loop Martin can rapidly flip through all the options open to him while contemplating the market, discard those that fill him with a momentary dread, and choose the one that feels just right.

These physical echoes of our thoughts are gut feelings, and we all, whether athlete, investor, firefighter or police officer, rely on them. I learned this basic piece of neuroscience the hard way. While trading on Wall Street, I often conceived trades that I thought were brilliant, identifying some securities that were cheap, others that were expensive. But my boss, habitually sceptical, would always ask, 'If the trade is so compelling and the money-making opportunity so amazing, why haven't other people spotted the trade? Why is the price discrepancy just sitting there on the screens for all to see, like a $20 bill lying on the sidewalk?' These were irritating questions, but in time I recognised their wisdom. For more often than not, trades conceived with obvious lines of reasoning turned out to lose money. It was a troubling discovery. Troubling because these trade ideas were usually arrived at using my best analytical efforts, drawing on my education and a wide reading of economic reports and statistics. I was acting as rational economic man.

In time, though, I realised I needed more than these cognitive operations. Often, while looking at a problem face-on and coming to some obvious solution I would catch a glimpse with peripheral vision of another possibility, another path into the future. It showed up as a mere blip in my consciousness, a momentary tug on my attention, but it was a flash of insight coupled with a gut feeling that gave it the imprimatur of the highly probable. An experienced trader, I think, learns to recognise these voices speaking from the fringes of awareness. To trade well you have to tear your attention away – and it can take a great deal of discipline – from the obvious piece of analysis lying under your nose, and listen to these faint voices.

LISTENING TO OUR BODY

And what an angelic choir they make. If only we could hear their music loudly and clearly, we would have at our disposal some of the most valuable signals in all the financial markets. For our bodies and pre-conscious parts of the brain, both cortical and sub-cortical, act as large and sensitive parabolic reflectors, registering a wealth of predictive information. They remain the most sensitive and sophisticated black boxes ever designed. When correlations between assets break down, when new correlations emerge, chances are our muscles, heart rate and blood pressure will register the changes before our conscious awareness. Just as galvanic skin conductance in the Iowa Gambling Task spikes before someone chooses cards from the losing decks, so too the bodies of experienced traders snap to attention well before they consciously understand the risks they are contemplating. Bodily signals, running ahead of consciousness, cry a warning. Yet traders often fail to heed them, because these messages are notoriously, frustratingly hard to hear. They fade in and out like a radio picking up a distant station, and leave us hanging on their every note, or worse, over-interpreting a burst of static. Our bodies and the pre-conscious regions of the brain may hear these songs clearly, and know what to do with the information they carry, but our conscious brain has only the most tenuous access to them.

111

In fact our conscious brain has surprisingly little grasp of what makes us decide to do one thing rather than another. A telling example of this ignorance has been provided by Joe LeDoux and Michael Gazzaniga, two neuroscientists who conducted a study of patients with a severed corpus callosum, the bundle of nerve fibres connecting the two hemispheres of the brain, leaving the two sides of the brain unable to communicate with each other. LeDoux and Gazzaniga gave instructions to these patients, via their right hemisphere (hemispheres can be targeted with instructions shown to either the left or right visual field), to giggle or wave a hand, then asked them, via the left hemisphere, why they were laughing or waving. The patients' left hemisphere had no knowledge of the instructions given to their right hemisphere, but the patients would nonetheless venture an explanation, saying that they were laughing because the doctors looked so funny, or waving because they thought they saw a friend. However implausible the answer, the patients were convinced they knew why they were acting in the way they were; but they were deluded in thinking so. Their self-understanding was pure confabulation.

A host of similar experiments have been conducted by Timothy Wilson and are reported in his book *Strangers to Ourselves*. He, like LeDoux and Gazzaniga, has found that people constantly trick themselves into thinking they understand the true springs of their actions. But the commentary people provide on their behaviour is often a meaningless accompaniment to action taken by pre-conscious parts of the brain. LeDoux, puzzling over his own and Wilson's observations, concluded that 'people normally do all sorts of things for reasons they are not consciously aware (because the behaviour is produced by brain systems that operate unconsciously) and that one of the main jobs of consciousness is to keep our life tied together into a coherent story, a self concept'. In other words, we make things up.

I found a disturbingly similar result in an experiment I conducted with a group of traders. I and a colleague were trying to find out how stress hormones respond to losing money and to high volatility in the markets. What we found was exactly what one would expect on the basis of previous stress research: the traders' stress hormones were remarkably sensitive to uncontrollability in their trading results, and to uncertainty and volatility in the market. So far, so good. However, in addition to the hard data I collected, in other words the physiological markers and the financial data, I also gave the traders a questionnaire to fill in at the end of each day, designed to determine among other things how stressed they were. What I found was that their opinions on how stressed they were had little if anything to do with reality, nothing to do with the fact that they might be losing money, or that their trading results seemed more than usually uncontrollable, or with market uncertainty as measured by its volatility. In fact, their opinions had little to do with anything I could discern. They seemed about as random, and irrelevant, as the confabulations provided by the split-brain patients. This result conjured up an almost comical picture of humans parroting words with little meaning and little connection to the physiological processes truly controlling their actions.

Odd as this finding may sound, it is a pretty standard one in endocrinology – opinions and physiology frequently travel along different tracks. What was odd was that the traders' hormones seemed to register risk far more accurately than did their opinions. Do their hormone-producing glands have a firmer grasp of financial risks than their frontal cortex? They could. If the traders were drawing heavily on pre-conscious processing of patterns coupled with gut feelings, they could display a disconnect between money-making skill and self understanding. This too is a pretty standard finding on a trading floor: it is frequently said that if you want to know what traders think of the market, do not ask their opinions, look at what trades they put on.

PHYSIOLOGICAL COACHING

We frequently and often comically misinterpret our actions. Given this unfortunate fact, we can appreciate the necessity when making an important decision of obtaining a second opinion. This opinion can take several forms. One of the most valuable sources of a second opinion, one that brutally and coldly exposes the faults in our reasoning, is the use of statistics. Alternatively, another person working with you, a coach say, can also help improve your decision-making. An increasing number of professions are coming to use coaches, and they have lately been appearing on more and more trading floors.

There is yet another form this external observation can take, and that is the form of physiological monitoring devices. If our bodies provide a highly effective early-warning system for both danger and opportunity, and if gut feelings tap into a wealth of experience, and if, furthermore, these somatic markers are largely inaccessible to conscious inspection, then perhaps we can hack into them by means of an external listening device, such as an electronic monitor.

Physiological monitoring could help scientists answer a number of intriguing questions, such as: do some people have better gut feelings than others? There is very little research which could help us answer this question, but at the same time there seems no *prima facie* reason why some people could not have better hunches than others. Training, of course, is essential in building up a library of trading patterns and developing hunches worth listening to; but as with athletes, traders differ in their physical endowments. Some of them may naturally enjoy more sensitive interoceptive pathways, and these people I am tempted to call hunch athletes. To qualify for this title a person would need to generate powerful somatic markers.

But hunch athletes would need more than strong somatic markers, for these are not of much use in themselves if we remain unaware of them. Of equal importance to gauging gut feelings is the measure-ment of our awareness of the signals. Here we do have some research at hand. Several scientists have found that sensitivity to somatic markers can be measured by means of a test called heartbeat aware-ness. In this test, participants are asked to time their heartbeats, or to say whether or not they are synchronised with a repetitive tone. Experiments with heartbeat awareness have found this marker is a good proxy for visceral awareness. The experiments have also found that heartbeat awareness is lower in people who are overweight, almost as if the signals are being impeded. Perhaps this is one reason trading floors are populated with relatively fit people.

This research raises the possibility of using tests of interoceptive awareness as a recruitment tool, to be used alongside regular interviews and psychometric testing, to help spot risk-takers with good gut feelings.

Could we also monitor traders' gut feelings while they take risks? Today a range of monitors can record heart rate, pulse, respiratory cycle, galvanic skin conductance and so on, and do so non-invasively. In fact, physiological monitoring of traders has been suggested by the magazine *The Economist*, when reporting on a result that emerged from one of our studies, mentioned above, on hormones in traders. That study had found that when morning levels of testosterone in male traders were higher than average, the traders went on to make an above-average profit later that day. The reporter for *The Economist* suggested that managers should test their traders first thing in the morning, and if their biochemistry was not just right, they should send the traders home. It sounds far-fetched, but this practice is already common in sport. Many sports scientists monitor their athletes' physiology non-stop, and look for just these sorts of signs that they are either ready for an upcoming match or need more work. In fact, today we are in a position to do the physiological spot checks suggested by *The Economist*. Such physiological monitoring could perhaps help managers tell when traders have succumbed to the siren call of irrational exuberance or the despondency of irrational pessimism.

In the future we might even be able to articulate the specific messages carried by our interoceptive pathways. Our conscious brain may have difficulty doing so, but science can help by intercepting and interpreting these messages. Some day we will be able to listen to our bodies and the subconscious regions of our brains and heed their warnings. Physiological monitoring devices, along with the computer back-up mentioned in the previous chapter, may one day provide human traders with what amounts to a hardened ecto-skeleton that may help them fight against the machines that increasingly dominate the markets.

This type of monitoring may seem futuristic, but many people already engage in it. There is a rapidly growing movement for what is called 'self-quantifying', recording one's own vital signs as a way of cutting through folklore and advertising and pop psychology to our own hard data. People are increasingly using a range of monitors to identify where in their lives the stress is coming from, what causes a bad night's sleep, what workout delivers the best results. There are even now, in development, many everyday consumer products that can perform real-time health monitoring, such as contact lenses employing bionanotechnology that sample cholesterol, sodium and glucose levels in your tears and transmit this information to a computer. Scientists have proposed a new type of toilet that similarly diagnoses your health based on urine analysis, and toothbrushes that do much the same with saliva.

I see no reason why this sort of physiological monitoring, if it is useful and popular with the public, Olympic athletes and the military, should not find its way onto the trading floor. And it is to the trading floor that we now return, in order to see how the physiology of risk-taking we have surveyed works in practice.

PART III

Seasons of the Market

FIVE

The Thrill of the Search

AN UNEXPECTED MESSAGE

After the brief excitement of the DuPont trade, the floor settles back into the lazy state that has been its lot for the past few days. Martin strolls back from the coffee room and hears nothing on the squawk and sees no hurried movement on any of the trading or sales desks, so his body receives the all-clear and gears down a final notch, returning heart rate and metabolism to a slow idle. The adrenalin dissipates. His vagus nerve gently takes control, and like a mother's hand on a troubled brow, it smooths away the last ripples of his bodily storm. The half a million dollars Martin has made trickles through his veins like some potent muscle relaxer. An inner glow of peace, goodwill and quiet confidence kindles and radiates. Money can do that.

The mild stress Martin has just experienced has been good for him, because it taxed both body and brain. This sort of effort is just what we are designed to do, so it makes for a satisfying experience. Effort, risk, stress, fear, even pain in moderate doses, are, or should be, our natural state. But just as important, just as vital to our health, the key to continued growth, is what sports physiologists refer to as the recovery period. Once a challenge ends, the fight-or-flight mechanisms should be switched off quickly, for they are metabolically expensive, and the rest-and-digest systems switched on. These recovery periods act much like a good night's sleep; but unlike the full eight hours doctors recommend, they are typically brief and frequent, like the short breaks in a boxing or tennis match. But no matter how brief, our bodies take advantage of the downtime to rest and repair, and over time these mini-breaks can add up to a healthy body and brain. Should we be denied these downtimes, even very brief ones, even when things are going well, our biology can become unbalanced, leading us into pathological mental and physical states and inappropriate behaviour. Such can happen on Wall Street.

Challenge, recovery, challenge, recovery – that is what toughens us. And that is why this trade has been good for Martin. He has benefited from just such a pattern of stress and recovery and has emerged a stronger – and indeed a richer – person. At this very moment, throughout his body, in a million different war zones, microscopic surgeons and nurses go to work repairing damage to tissues, tending to his every comfort – and brother, does it feel good.

Martin turns down the aisle that leads deep into the hinterland of trading and sales desks, the grand trunk of the trading floor. Normally a speedway of frenzied bankers, today it feels more like the main street of a small town. As he enters the corporate bond department, one of the traders, puzzling over what looks like a credit card statement, looks up and nods his regards. A frisky salesman shadow-boxes as he passes. Walking by the arbitrage desk, Martin intercepts a tennis ball thrown by Logan to Scott. He tosses the ball to Scott, who tells him the brokers are having sushi sent up for lunch. Back at the Treasury desk, located between the arbitrage and mortgage desks, Martin casts an affectionate glance across a trading floor that has given him so much, and listens to its familiar and reassuring sounds.

Martin decides to treat himself to a luxury that is rare on Wall Street – reading parts of the newspaper other than the business section. He puts his feet up on his desk and with satisfaction opens the paper at the arts and review section. Someone down the aisle announces they have extra donuts; a woman on a far-off sales desk lets out an occasional high laugh.

Martin relishes the lazy hours that stretch ahead, but anyone watching him would, after a while, notice him hesitate, consider. As he glances at the screens, a slight tension creases his face and he shifts uncomfortably in his chair. Unbeknownst to Martin's conscious brain, a subsonic tremor has just shaken the market, and silent shock waves radiate from the screens, reverberating in the cavern of his body. Something is not right. The screens flicker at a different frequency, the matrix of prices dance into a new pattern, like a single turn of a kaleidoscope. Volatility has hardly budged, but the minuscule changes are unexpected, and nothing snaps us to attention faster than the unexpected, something novel emerging out of an indifferent background.

Martin, an Olympic-class hunch athlete, is often the first to sense these things, but others are not far behind. All across the floor the inaudible call from the market receives its echo in the bodies of traders and salespeople. For some of them, muscles tense ever so slightly; for others, pupils dilate and breaths come a bit faster; for still others, stomachs tense and hunger abates. An observer might notice postures straighten, conversations become more animated, hand gestures more abrupt. Few people are yet aware of the changes taking place in their bodies, but the cumulative effect is much like someone turning up the volume on the trading floor. A good manager should sense the budding commotion, see the restlessness. And now, like some large beast stirring from a deep slumber, the trading floor returns to life.

THE MARKET'S MORSE CODE

What was this shock that emanated from the screens? What was it that vibrated pre-consciously in the taut membrane of Martin's early-warning system? That shock was information, and information manifests itself in the shape of novelty. When the world sends us a message it does so through the language of surprise and discrepancy; and our ears have been tuned to its cadences. There is nothing that fascinates us more, little that agitates the body more completely. Information warns us of danger, prepares us for action, helps us survive. And it enables us to perform that most magical of all tricks – predicting the future.

The link between information and novelty was discovered and brilliantly explained by Claude Shannon, an engineer working at Bell Labs in the 1950s. According to Shannon, the amount of information contained in a signal is proportional to the amount of novelty – or, put another way, the amount of uncertainty – in it. That may seem counter-intuitive. Uncertainty seems the antithesis of information. But what Shannon meant was this: real information should tell us something we do not already know; it should therefore be unpredictable.

Most messages we encounter in everyday life are, however, predictable. Usually we sort of know what is coming next when we read a book or hear someone talk, because most messages contain a lot of noise – that is, words or characters which could be compressed out of the message without impairing its meaning. People composing text messages accordingly condense the sentences they want to send, just as people did in the old days when using telegraphs, to eliminate any characters or words that could be predicted, leaving behind only what could not be predicted, the true information content of the message. For example, imagine sending the following text message half an hour after you were expected home: 'I am late. The car has a flat tire. I will be home in one hour.' This message, 63 characters long including spaces, contains a lot of redundancy, and can be compressed. To begin with, if you were due home half an hour ago, obviously you are late, so the first sentence can be cut. And obviously it is the car that has a flat tire, what else could it be? So you can drop the reference to it. And obviously it is you who will be home, so the pronoun can be implied rather than stated. Eliminating the redundancy, you send instead: 'Flat tire. Home 1hr.' This message, 20 characters long, has been compressed so that it contains only the information your family could not have predicted. If they were to receive it word by word, they could not guess what word would come next. This simple example illustrates the fundamental discovery of Shannon's information theory: information is synonymous with unpredictability, with novelty. When receiving pure information we are in a state of maximum uncertainty about what comes next.

Our sensory apparatus has been designed to attend almost exclusively to information. It ignores predictable events but orients rapidly to novel ones. The cerebellum provides a nice illustration of this principle. When we plan an action our neo-cortex sends a copy of this plan to the cerebellum, which then dampens or even cancels out the sensation it expects to result. Because of this dampening we are largely unaware of, say, our arms moving back and forth when we walk, or the chafing of our own clothes on our skin. It is also the reason we are unable to tickle ourselves: since we have produced the motion of fingers on ribcage, our cerebellum dampens the expected sensation; we may still feel our fingers on our skin, but we are not surprised, so the tickling has no effect. Why would we want to suppress sensations we expect to come from our own actions? Because doing so proves extraordinarily useful in a control mechanism: if the sensory feedback from an action is exactly what we expect, we do not need to pay attention to it. If however the feedback is other than what we expect, then it carries information: something has gone wrong with our plans, and this information teaches us to calibrate our movements to our intentions.

An extraordinary further illustration of the principle that we attend largely to the unexpected can be found by considering the visual system of the common frog. Evidence suggests that frogs are blind unless something moves in their visual field. Frogs do not, apparently, have any interest in gazing out on their pond just to appreciate its beauty; their blindsight registers objects only when movement indicates the presence of an insect to eat or a threat to escape. The frog's eye thus presents a pure example of a sensory system doing just what it was built to do – attending exclusively to information.

Human sensory systems work in much the same way. We too lose sight of objects if they do not move, an effect known as Troxler fading, after a nineteenth-century German physiologist who noted that we gradually lose awareness of unchanging visual stimuli, just as we do the constant sound of background traffic. However, we rarely notice fading such as the frog-eye effect, because we move our eyes and head almost continuously, and this makes our visual field move. But you can experience something like it if you have someone hold their hand out to the side of your head, so it occupies the very edge of your peripheral vision. When their hand is motionless you will not see it, yet when they move it you will. This example points to yet another problem – in addition to those surveyed in Chapter 3 – with the commonsense notion that our senses operate like a movie camera, recording non-stop the sights and sounds around us. Troxler fading shows that our senses do not work at all like that. Indeed, it is probably closer to the truth to say that we, like the frog, are built to ignore the world unless something of importance happens.

Such a sensory system admirably reduces demands on our attentional resources, but in the modern world it can also lead to problems. Yes, we are built to attend to novelty, but unfortunately we do not function particularly well in its absence. Without it we can suffer stimulus hunger, and this can lead to a condition among drivers (even, it is claimed, among pilots of commercial aircraft) which would be comical were it not occasionally dangerous, a condition variously called 'highway hypnosis' or 'the moth effect'. Drivers on long, featureless stretches of road or driving through the night can become so starved for stimulation that they attend almost hypnotically to the rare appearance of a light beside the road, often a police car with lights flashing, and then proceed to drive straight into it.

Information thus holds a strange and powerful attraction for us. When in its presence, we come alive. Entering your home and noticing the furniture out of place; hiking through the woods and hearing the crunch of twigs behind you; reading a mystery novel and realising in a creepy moment that the hero has just used the same turn of phrase as the murderous psycho the police are hunting. In these situations your awareness sharpens and your attention zooms in on the unexpected scene – 'What the hell!' your pre-conscious brain utters, and in that very instant your world morphs from an indifferent and impressionistic background into a scene of hyperrealism.

The mechanism operating in your brain at this point is a marvel of chemical and electrical engineering. When the alarm centre of your brain is tripped, neurons in the locus ceruleus, located in the brain stem, boost their firing rate and spray a neuromodulator called noradrenalin throughout your brain (see fig. 8). Neuromodulators are a type of neurotransmitter – the chemicals used to bridge the synaptic gap between neurons so an electrical message can jump from one to the other – but of a very particular kind. They do not participate in any specific brain activity, like doing maths or speaking French or remembering the dates of the Punic Wars; rather they alter the sensitivity of neurons throughout the brain, making them fire more easily or more rapidly. The effect noradrenalin has on neurons can be compared to that of turning up the lights in a room and the volume on a microphone.

That is what is happening to Martin right now. His early-warning system has sprayed noradrenalin throughout his brain, bringing him to a state of high alert, enhancing arousal and vigilance and lowering sensory thresholds, so that his senses are put on edge, enabling him to hear the faintest sound, notice the slightest movement. Reaching the neo-cortex, the noradrenalin also improves the signal-to-noise ratio of incoming sensory data. This is an extremely useful trick. When in a relaxed state Martin scans his environment randomly and widely, just as a radar sweeps 360 degrees, and a low signal-to-noise ratio is to be expected. But when he is surprised by an unexpected event, as he is now, his senses are drawn to a focal point, he filters out background sensations and concentrates instead only on that information relevant to the problem at hand. This radar-enhancing property of the locus ceruleus is partly responsible for what is known as the Cocktail Party Effect, our occasional ability to pick out a voice on the other side of a crowded room. Animals on the hunt, athletes in the heat of competition and traders making money rely on this focused attention and these supernatural senses. As do soldiers in the field: 'The moment that the first shells whistle over and the air is rent with the explosions,' explains Erich Maria Remarque, 'there is suddenly in our veins, in our hands, in our eyes, a tense waiting, a watching, a heightening alertness, a strange sharpening of the senses. The body with one bound is in full readiness.'

The locus ceruleus thus arouses Martin's brain and, crucially, also his body. It projects its neuronal fibres up into higher reaches of the brain and down through the fight-or-flight nervous system into the body. Here it sprays noradrenalin onto tissues in the heart, arteries, bronchial tubes and adrenal glands (see fig. 8). It brings his body to a state of preparedness, so that once his brain has figured out what threat looms and what action is required, his body is ready to initiate it. The information the locus ceruleus records is of a low quality;

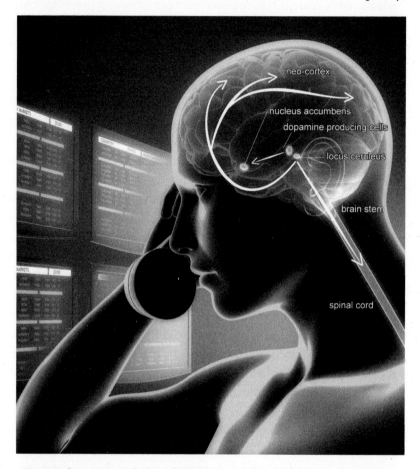

Fig. 8. Information and arousal. The locus ceruleus projects noradrenalin up into higher regions of the brain where it makes our senses more acute, and raises the signal-to-noise ratio of incoming information so we can focus on a current threat or opportunity. It also projects down into the body where it triggers the fight-or-flight response. Dopamine-producing cells in the brain stem project to the basal ganglia; one of its target areas here is the nucleus accumbens, often called the thrill centre of the brain. Dopamine encourages us to take risk, to engage in physical activities, like hunting, foraging and trading, that lead to uncertain rewards.

unarticulated, it tells Martin little more than, 'Pay attention, something's up!' The information may be low-grade, but its transmission is fast, and therein lies its value. When a correlation between events breaks down or a new pattern emerges, when something is just not right, chances are it is the locus ceruleus that responds to the change long before conscious awareness. And it trips a very basic alarm, preparing us for fast reactions. It pulls taut the membranes of our recording devices, kindles the fire of our metabolism, and cocks our muscles, placing them on a hair trigger.

Through its influence on the locus ceruleus, information thus registers as a lot more than mere data: it registers as a bodily reaction. Information becomes physical. So tight is this relationship that Daniel Berlyne, a psychologist working at the University of Toronto in the 1960s, graphically plotted arousal against information and found an elegant ∩ shape. What Berlyne's hill tells us is that low levels of information, such as we encounter in a dull conversation, leave us bored and sleepy, while high levels of complexity, such as we might find in a movie with a difficult-to-follow plot or in an overload of files at work, confuse us and promote a state of anxiety. But just the right amount of information piques our curiosity, quenches our thirst for novelty, and provides a diffuse feeling of satisfaction that spreads throughout body and brain.

To avoid swamping us in complexity and creating constant anxiety, the brain must distinguish significant information from the trivial. The locus ceruleus cannot do this on its own. To sift the meaningful from the meaningless, we rely on visceral judgements not unlike gut feelings. Several higher-brain regions, in the hippocampus and posterior parts of the neo-cortex, yet still operating below the level of conscious awareness, scroll rapidly through patterns stored in our memory banks and compare them with the facts now before us. The process of pattern-matching is given an urgency and a motivational edge by the amygdala, which tags each pattern with an emotional feeling, giving you a quick and dirty assessment of the potential threat or opportunity facing you. This tag team of pattern recognition, emotional assessment and alarm centre sifts information and provides us with the gut feelings we need to prevent us from being fooled by unimportant information.

We feel these fluctuating assessments of importance throughout the day. Some event catches our eye, and we hover between attending to it or turning away. Noticing a chair unexpectedly hanging from the ceiling of an art gallery may provoke a moment of curiosity, but unless the installation has some deeper significance we quickly habituate to the scene and lose interest. In fact, without the guiding hand of our gut feelings, the locus ceruleus would be forever lost in wonderland, distracted, like a child, by ever new sights. A lot of fantasy literature relies on just this sort of initial amazement. But what separates the throwaway books of this genre from classics such as Ursula LeGuin's *Earthsea* novels is a parable that taps into our deepest concerns. The visual arts today also rely on what the critic Robert Hughes calls the Shock of the New, but here as well our gut feelings distinguish (even if the market does not) between the deep and the shallow. The locus ceruleus may be tricked by shocking art, just as it is by eye-catching advertising; but only if the amygdala and higher-brain regions, and indeed our entire body, are engaged does a surprising piece of art, like the best conceptual installations, find roots in our deepest uncertainties and promote a satisfying and lingering arousal. Good art critics rely on gut feelings just as much as profitable traders.

The need to distinguish trivial from important information is nowhere more pressing than in the financial markets. Every newsworthy event across the globe, be it the Bank of Japan raising interest rates, the announcement of Chinese industrial production, Eurozone inflation, a hurricane approaching the Gulf of Mexico, shows up on scrolling news feeds and in market prices. The information pours in non-stop, like a never-ending and uncompressible message, a telegraph that never stops clicking. Every change in the world shows up as price changes in the bond, stock, currency or commodity markets. The more information pouring in, the greater the uncertainty, the higher the market volatility. Financial market volatility thus provides the most sensitive barometer of what one might call global arousal, the amount of novelty in the world. In fact there is a futures contract traded on the Chicago Board Options Exchange called the VIX which tracks this very uncertainty, being an index of the financial community's expectation of how much the market will move in the coming months. It has for good reason been called 'the Fear Index', and when the credit crisis of 2007–08 erupted the VIX spiked from a sleepy 12 per cent to over 80 per cent in a matter of months.

A human brain trying to map all the information in the financial markets would soon collapse with exhaustion. Few professions, with the exception perhaps of air-traffic control or the military during time of war, compare with finance for the amount of information that must be sifted and processed in real time. But skilled traders and investors can do it. They can separate the signal from the noise, and feel in their bodies when the chaos on the screens can be safely ignored and when it cries a warning that should be heeded. Good traders like Martin do not just process information, they feel it. There are few phenomena in finance more remarkable, even mysterious, than this close linkage between market and body.

The picture conjured up by this research on information and arousal is of a constant call and echo between market and trader. The market broadcasts its information, beats out its message on the tom tom of volatility, and the very body of a trader, like a tuning fork, vibrates in sympathy. I am not sure traders are ever fully aware of this fact, since much of it takes place pre-consciously. Judging from my own experiences, my observations of other traders, and my experiments on a trading floor, I would say they usually are not. Yet market information and trader arousal wax and wane together, dragging the trading floor, whether it is aware of it or not, willing or not, across Berlyne's ∩-shaped hill, from boredom through excitement to anxiety and stress.

THE FED!

And that is why today, at around 11 a.m., the first inklings of the coming storm were felt in the bodies of traders and salespeople. The bodies of these unsuspecting bankers have already taken the first steps in gearing up their defences. Arousal flickers and their gaze orients to the disturbance. Logan, in mid-throw, looks over his shoulder at the screens. Scott has already wheeled his chair back to his desk. They sense something is wrong, but are not sure what. One by one, on desk after desk, and all along Wall Street, traders and salespeople dip their newspapers to look at the screens, phone conversations are politely cut short – 'Look, can I call you back?' – donuts remain half-eaten. Martin, Gwen, Logan and Scott, electrified by heightened senses, quickly register the slight changes taking place on the broker screens and begin the cognitive task of figuring out what is agitating the market.

Ash, the head of the trading floor, looks up from his papers and steps out of his office, surveying the floor. Then, strolling up to Martin, hands in pockets, he asks, 'What's up?'

Martin, twirling his pen and holding the screens in a steady gaze, replies, 'Not sure. Feels like the market might break down.'

At that point a salesman on the mortgage desk yells to Martin, 'Wells Fargo hearing the Fed may raise half a point this afternoon. You know anything about this?'

Martin and Ash look momentarily shaken. But Ash quickly dismisses the suggestion. The Fed does not leak rate moves in this way. If it wants to warn the markets, and it usually does, it hints at its intention of changing interest rates weeks in advance, not on the very day when the Board of Governors meets to finalise its decision. Most bankers know this, yet the market appears to be taking this rumour seriously. As the rumour mill of Wall Street goes to work refining the story, it emerges that one of the governors of the Fed gave a speech last night to a small group of senior bankers in which he spoke in no uncertain terms about how concerned the Fed is by what it considers an unjustified rally in stocks. He made it known that the Fed will not tolerate a bubble and the threat it poses to the stability of the financial system. One of the bankers in the audience took the speech as a pretty clear message that the Fed would raise rates today, and after giving his own bank time to set its positions for the rate hike he passed on his views to clients. Hence the news now seeping into the market.

An increase in interest rates by the Fed, especially on a day when no one expects it, would send a tidal wave of volatility through all financial markets, devastating prices. The interest rate set by the Fed acts as the benchmark against which all assets are valued, so when it changes the prices of all other assets have to change as well. Assume that the Fed has set its interest rate at 5 per cent. And further, assume that you hold your savings in a balanced portfolio of assets: you own stocks which pay a dividend of about 3 per cent and which in addition go up in price on average about 4 per cent a year; bonds which yield 5 per cent; and a small amount of commodities, such as gold, which yield nothing but go up in value during times of inflation. Now, what happens if the Fed raises its rate to 6 per cent? All of a sudden the assets you own no longer look so attractive. Your stocks now yield a full 3 per cent less than what you could earn in a savings account; your 5 per cent return on bonds looks paltry compared to the 6 per cent you could receive on new bonds; and the money you have invested in gold, forgoing the 5 per cent you could have received from bonds, could now be earning 6 per cent, making it that much more painful to hold this inert metal. Consequently, rate hikes by the Fed commonly depress the values of all assets. That is why experienced hands on Wall Street keep in mind a sage piece of market lore: never fight the Fed!

This market is not in a fighting mood, so it acts accordingly. Stocks sober up to the reality that their party may be coming to an end. The Fed could take away the punchbowl of easy money, and if it were to do so in earnest it would spell the end of the glorious bull market that has lifted stocks almost 40 per cent over the past two years; so in the next half hour the S&P index drops almost 2 per cent. Commodities too take a hit, with gold down $5 an ounce and oil $2 a barrel. However, it is in the bond market, the market for interest rates itself, that the news has its most immediate and undiluted effect. If the Fed were really to raise rates half a per cent, the bond market would be knocked to its knees.

Ash asks how the desk is positioned, and is relieved to hear that it has no large positions, nothing dangerous. He rushes off to talk to other trading desks. Martin calls his traders together for a hasty meeting. A dozen traders lean in to the pow wow. What happens to bonds, they speculate, if the Fed hikes rates a quarter point? Half a point? What happens to stocks? But their strategising is cut short. The squawk box begins to crackle with salespeople around the world calling in client trades – bid $150 million five years for Industrial Bank of Japan, $375 ten years for Monetary Authority of Singapore, $275 million long bonds for a French asset manager – and the Treasury market begins to fall. Martin has been caught unprepared for this avalanche of business and the yawning hole that has opened up in the Treasury market, taking the market down a quick half-point. Quickly adapting to the volatility and to the non-stop client trades, he disappears into his zone; this is what he has trained to do, and he does it well.

The initial trades he parries well, as does Gwen, buying bonds from clients and selling them into the screens, making small amounts of money on some but mostly breaking even, which in this terrified market is a relief. In the last fifteen minutes, however, the client selling has turned relentless, pushing the ten-years' total loss to almost a full dollar. Even an old hand like Martin has trouble staying one step ahead of the market. The last block of bonds he bought from a client he has not managed to scratch, having to sell half of them at levels below where he bought them, and the other half he has not been able to sell at all. As the market continues to fall, these bonds start losing money at an alarming rate. Other traders along the desk find themselves in a similar predicament, and slowly retreat behind a barrage of selling. Gwen struggles with $450 million fives, most of which she bought from the Caisse de Depot, a government fund in Quebec. Martin has to concentrate on getting out of the desk's existing risk rather than buying more bonds, so he lowers the bids he shows clients and starts missing business. Traders are always caught between the need to make money and the need to keep clients and sales force happy. In markets heading in one direction, and fast, as they are today, the two usually conflict. As the morning wears on, the Treasury desk leaves a trail of disgruntled salespeople, and the goodwill established by the DuPont trade quickly burns off.

Then, as these things generally go, the rumours, never terribly reliable in the first place, get hijacked by fear, and reason deserts the field. Rumour now has it that the Fed will hike rates three quarters of a point, and that this may be just the beginning of a concerted round of tightening to follow in the months ahead. The sell-off turns into a freefall, the ten-year dropping another half-point without much trading taking place, the five-year about the same. In fifteen minutes Martin has lost maybe $1.75 million on the bonds he has not been able to sell, Gwen another $2 million on her five-year position. The Treasury traders now feel besieged. Altogether they have lost well over $4 million on long positions they cannot get out of, a lot more than Martin made on the DuPont trade, and a murmurous concern spreads among neighbouring desks. What is more, black boxes, taking advantage of the fear and volatility and the traders' loss limits, are pushing the market down, hoping to make traders panic.

For Martin, this sell-off has turned into a white-knuckle ride. But he is a hardened campaigner from much worse crises, and he does not lose his cool. Immersed in the flow and absorbed by the activity, he nonetheless mulls the information, scrutinising every price change, its size, speed, the volume of bonds traded; he listens to the client trades streaming onto the floor, through both his desk and distant ones – the mortgage desk, the corporate – and in the back of his mind he queries the rumour, whether it is credible, whether the market is reacting correctly even if it is true; and at a still more abstract level he sifts recent economic statistics to get a sense of the macro-economic reality, asking himself, is this economy strong enough to withstand higher interest rates? Layer upon layer of brain churns the data, searching for a pattern that feels right.

And then each layer of analysis turns up a match, like spinning wheels in a slot machine coming to rest one by one on a single fruit. One-two-three. Martin feels a gestalt-like switch in his body; the knot in his stomach unties and he senses a budding confidence. He has a hunch. A new interpretation tugs from the periphery of his consciousness, a mere glint of a possibility. No way this economy can take a large rate hike, let alone a series of hikes. No way this market should be scared even if there is a hike today, because it would reduce the chances of inflation, the bond market's greatest enemy. Besides, this last move down and the desperate selling that drove it there felt like a last gasp, the panic move known in the market lexicon as the 'capitulation trade', when nervous managers, terrified of further losses, want out of their bonds at any price. When fear dominates, cool heads step in, and Martin is one of them. But he is not alone. In the past five minutes the market has started to trade differently. It is still pounded by large selling, in fact nothing but selling, and prices continue to spike down, but they bounce back quickly, like a trampoline. Something big, invisible, lurks in the depths, some immense client stepping in to buy every time the market dips, but who? Maybe Bank of China, maybe Bank of Japan, maybe the Kuwait Monetary Authority, who knows? But whoever it is, this is one large beast. Gwen has felt it too, and together she and Martin decide to hold any more long positions they buy from clients.

They start to see some buying, clients stepping in to nibble at the market. Before long it becomes apparent that the panic move down is over, and the market has returned to a normal battle between buyers and sellers. It is in these conditions that traders can make a fortune. Volatility is high, client volumes are large, and since the market is jumping around so much and uncertainty is so high, clients are not too demanding about the prices they receive. They want fast execution, and are willing to leave money on the table to get it. For example, in one trade, with the five-year bond trading at 100.16–18, an insurance company in Florida asks to sell $80 million, and Gwen bids 100.14 and buys them; a few minutes later another client asks to buy $100 million and she offers at 100.19 and gets lifted, making a quick $125,000. On a day like today this seesaw between clients buying and selling can go on for hours at time. Profit margins may be small, a cent here, half a cent there, but with the volumes across a bank amounting to tens, even hundreds of billions of dollars, they add up. The mighty edifice that is Wall Street was not built on the fortunes of flamboyant speculators, as myth would have it – it was built on pennies.

Just as valuable as the bid–offer spread the traders enjoy is the information they alone can access. Traders at the big banks see the largest client trades, and therefore know before the rest of the financial community where the smart money, the big money – the central banks, the hedge funds, oil money, large pension funds, the sovereign wealth funds – is going. This enables them to get there first. They can sell large blocks of bonds to their clients at market prices, and then buy them on the screens before the smaller banks and institutions figure out what is going on and why they are losing money. No one else in the world is privy to this fund of information, although the traders do share it with their big clients – especially the hedge funds they hope to work for one day – and so valuable is it that the big banks spend fortunes maintaining an extensive global sales force.

More and more traders across the floor, seeing buyers emerge, have adopted the same strategy as Martin and Gwen, building up a long position in bonds. Martin senses relief along the aisle. The traders are no longer battling for survival; they are warming to their game. Like a hockey team that has scored an important goal, the traders and salespeople feel a shift in spirit from defence to offence. This is the type of market traders dream of, one that draws them out of their fears and thoughts and preconceived ideas. Martin, Gwen, Logan, Scott and the other traders along the aisle no longer view the market's volatility as a threat to fear, but as a challenge to embrace. They have entered the zone. When they are in this state, the pulses of information being shot at traders are welcome, the risks gladly embraced, the uncertainty craved like an exciting game. Here, at the peak of Berlyne's hill, information comes freighted with excited expectation.

THE PLEASURE OF INFORMATION

We are so completely enthralled by information that one could, without exaggeration, say we are addicted to it. The addiction develops under the influence of another neuromodulator, this one called dopamine. Produced by a group of cells at the top of the brain stem, dopamine targets brain regions controlling reward and movement. When we receive some valuable piece of information, or perform some act that promotes our health and survival, such as eating, drinking, having sex or making large amounts of money, dopamine is released along what are called the pleasure pathways of the brain, providing us with a rewarding, even euphoric, experience. In fact our brain seems to value the dopamine more than the food or drink or sex itself. Give an animal the choice between on the one hand eating and drinking, and on the other self-stimulating with dopamine, and it will self-stimulate until it starves. If noradrenalin modulates the brain's overall level of arousal, how awake and attentive it is, dopamine modulates its level of motivation, how eagerly it wants things.

Unfortunately, dopamine neurons are easily duped, and can be tricked into bestowing their rewards by drugs of abuse. Almost every recreational drug, be it alcohol, cocaine or amphetamine, achieves its addictive effects by increasing the action of dopamine in a region of the brain called the basal ganglia, located midway between the brain stem and the cortex, and specifically in one part of it called the nucleus accumbens (fig. 8). If we view dopamine as the normal compensation we receive for valuable effort, then recreational drugs are in effect running a scam, tricking our brain into paying for healthy activities we never actually performed. To get an idea of just how effective this scam can be, consider the numbers. Food can raise an animal's dopamine levels by 50 per cent, sex by 100 per cent. However, nicotine can raise them by 200 per cent, cocaine by 400 per cent, and amphetamine by 1,000 per cent. Give junkies the choice between food and self-stimulating with dopamine, and not surprisingly they too lose interest in eating.

It is tempting to conclude that dopamine is the molecule of pleasure, but unfortunately things are not quite that simple. When scientists tested this idea they found something they had not expected. If they gave an animal a shot of juice, say, they found that it experienced a spike in dopamine, just as you would expect if dopamine was coding for the pleasure of drinking. So far so good. But after giving the animal several more mouthfuls of juice they found something odd happening – the shot of dopamine in the animal's brain started drifting forward in time, so that it actually occurred before the animal drank. The dopamine spike came to coincide with the appearance of cues, a sound perhaps, or an image, that reliably preceded the consumption of the juice. Put another way, the dopamine spiked when the animal received information predicting the imminent arrival of pleasure.

How, the scientists wondered, could an animal get pleasure before actually drinking? Some of them started to suspect that maybe there are two different types of reward – the pleasure of consumption and the pleasure of anticipation – and that dopamine has more to do with the latter. Other chemicals in the brain, such as natural opioids, may provide the pleasure of actual drinking, but perhaps dopamine provides something that is closer to a desire, even a craving. Desire is more of an anticipatory feeling; but is nonetheless powerfully motivating and in some sense enjoyable, although at times it can be more like a maddening itch. Two of the scientists conducting this pathbreaking research, Kent Berridge and Terry Robinson, concluded that dopamine stimulates the *wanting* of juice rather than the *liking* of it.

In humans, dopamine works in much the same way, causing us to value cues that predict pleasure, cues such as the smell of our favourite restaurant, the exciting appearance off in the distance of ski slopes, or a certain form-fitting blue sweater worn on a date. Seen in this light, perhaps it is dopamine as well that drives our perennial obsession with money, the ultimate predictor of good times.

There is another wrinkle to the dopamine-as-craving story. Give a monkey a single squirt of juice and its brain shows a spike of dopamine, but repeat the process several times and eventually the dopamine neurons settle down. But now give the monkey two squirts when it expected one, and dopamine perks up once again. Give it three squirts and dopamine perks up even more. Yet if these three squirts are now repeated, dopamine once again settles back down. What this means is that the amount of dopamine released into the nucleus accumbens does not depend on the absolute amount of reward an animal receives, but on how unexpected it is. This further suggests that we enjoy and crave environments in which we receive unexpected rewards; in other words, we enjoy risk. Put another way, dopamine spikes with information; and it acts as a learning signal, making us remember what we have just discovered. Some neuroscientists, such as Jon Horvitz at Columbia and Peter Redgrave at Sheffield, have even gone beyond the dopamine-as-predictor-of-pleasure idea and controversially argued that any experience, even an unpleasant one, that helps us predict future sources of pleasure and pain can deliver a shot of dopamine.

Dopamine research has changed the way psychiatrists understand and treat drug addiction. Medical researchers have found that the brain chemistry of people taking drugs evolves along the same path as that of animals receiving juice. The drugs first deliver a pleasurable hit and a potent shot of dopamine, but with increasing use the dopamine signal drifts forward in time and attaches to cues predicting the taking of drugs – certain music, or people, or special places, such as a nightclub – and these stimulate a near-irresistible hunger. The really powerful motivation is now the craving of the drug rather than the pleasure it provides. Many addicts actually come to lose the pleasure they once enjoyed from drugs, may even find the actual consumption distasteful, but cannot stop. Smokers cannot resist the temptation to smoke, but often find the act itself a nasty one, leaving them feeling terrible afterwards. In order to kick a habit they now find unpleasant, addicts often find they have to separate themselves from drug-taking cues by changing neighbourhoods and avoiding old friends. Many anti-drug advertising campaigns have backfired by misunderstanding this point. These campaigns often featured images depicting the horrors of addiction, maybe a bloody syringe and a dark alley; but these images were the very ones predicting the consumption of drugs, and therefore delivered a large dopamine hit in many reformed addicts, perversely renewing their craving and driving them back to heroin or cocaine.

What else besides drugs of abuse can create a dopamine-driven craving? If dopamine fuels a desire for information and unexpected reward, perhaps it also fills us with a burning curiosity. Perhaps curiosity itself, the need to know, is a form of addiction, making us race to the end of a good mystery novel, or driving scientists to work day and night until they discover insulin, say, or decode the structure of DNA, scientific breakthrough being the ultimate hit of information. When the Theory of General Relativity dawned on Einstein, he must have had the mother of all dopamine rushes.

Gambling, with its unexpected rewards, can also become a dopamine-driven addiction. Plugging coins into a slot machine hour after hour may look the epitome of boredom, but when those three fruits line up unexpectedly and you hear that waterfall of coins, large quantities of dopamine are released into your brain, leaving you with a craving for more. And if gambling can be addictive, why not trading? Trading provides some of the highest rewards available in our economy, but they are highly uncertain, and attaining them entails predicting the future and taking huge risks. So it may be dopamine that delivers the powerful high traders feel when their trades work out. It is no wonder that many observers suspect that traders on a roll may be in the grips of an addiction. And like an addict who quickly habituates to a given dose of a drug and has to continually increase the hit, traders too may habituate to certain levels of risk and profit, and be irresistibly compelled to up their position size beyond what would normally be considered prudent.

Importantly, dopamine, like noradrenalin, does a lot more than motivate the brain: it also prepares the body for action. In the words of Greg Berns, a neuroscientist from Emory University, 'In the real world, action and reward go together. Goodies don't just fall in your lap; you have to go out and find them.' And it is dopamine that drives this search. That is what one research group from Germany found when they designed an experiment to disentangle the pleasure of eating from the desire to search for food. They pharmacologically depleted dopamine in rats, and found that the animals would continue to eat and enjoy food if it was put directly into their mouths, but would not walk even a short distance to obtain it.

When we look at this connection between movement and reward we glimpse the very motivational core of our being, what thrills us, why we take risks, why we love life. For dopamine does a lot more than merely tag information with a hedonic colouring; it also rewards us for physical actions that lead to unexpected reward, such as trying a new and successful hunting technique or stumbling on a particularly rich patch of berries when foraging in the woods, and it makes us want to repeat these actions. Indeed, under the influence of dopamine we come to crave these physical activities. As Berns says, research into dopamine has turned 'upside down a basic tenet of economics', for much of it has found, somewhat counter-intuitively, that animals prefer to work for food than to receive it passively.

A preference for effortful consumption makes sense from an evolutionary point of view, for both animals and humans. If you are programming an animal to survive, you should make it enjoy more than just eating and drinking and having sex, which would encourage it to develop into nothing more than a couch potato or a louche hedonist. You should make it love the activities that lead to the discovery of food, water and sex. And that is what dopamine does, it makes us want to repeat certain actions, be they hunting, going on dates, or for that matter searching the screens for a trading opportunity. A clear statement of this principle can be found, improbably, in the film *Jurassic Park*. When a group of visitors watch across an electrified fence as a goat is tethered to a stake, lunch for the resident T. rex lurking somewhere out of sight, Sam Neill comments ominously that this predator 'doesn't want to be fed. He wants to hunt.'

If we pull together the various strands of research on dopamine, we could say the following: that dopamine surges most powerfully when we perform a novel physical action that leads to unexpected reward. Dopamine drives us to push beyond established routines and to try new search patterns and hunting techniques. As a result, the effects of dopamine on the course of evolution have been revolutionary. According to Fred Previc, a psychologist at Texas A&M University, the rapid growth of dopamine-producing cells, the result of ancient dietary changes such as an increase in the eating of meat, changed history. It encouraged us to take risks just for the hell of it, independently of any rational expectation of gain. Dopamine fuelled a robust lust for life, with all its vicissitudes. You may well imagine just how fateful a day it was on the African savannah when the new dopamine-driven brain was given the keys to the mammalian body, with its awesome metabolic resources, for then humans evolved into what they are today – voracious and marauding search engines, Google on wheels.

John Maynard Keynes, more than any other economist, understood these subterranean urges to explore, calling them 'animal spirits' – 'a spontaneous urge to action rather than inaction'. He considered them the pulsing heart of the economy. 'It is a characteristic of human nature,' he wrote, 'that a large proportion of our positive activities depend on spontaneous optimism rather than on a mathematical expectation.' Should this spontaneous optimism falter and animal spirits dim, leaving us with nothing but mathematical calculation, then, he warned, 'enterprise will fade and die'. He suspected that business enterprise is no more driven by the calculation of odds than is an expedition to the South Pole. Enterprise is driven to a great extent by a pure love of risk-taking.

It is a core principle of formal finance that higher returns come only by taking greater risk, and much the same can be said of our ancient search and hunting patterns. Dopamine prompted us to try things we had not tried before, and in so doing led us to stumble upon valuable territories and hunting techniques that otherwise would have remained undiscovered. It pushed us to venture past protective barriers. 'I have never been beyond the edge of the jungle. I wonder what it would be like in that wide-open savannah.' 'I wonder if a differently-shaped spear would work better.' 'I wonder what lies beyond the horizon?' Even though answering these questions entailed great danger and resulted in countless deaths (in a very real sense curiosity does occasionally kill the cat), it nonetheless proved of great value in our long history of geographic, scientific and, yes, financial exploration. Dopamine, we could say, is the history molecule.

Mysteries remain in this intriguing molecule, and one in particular comes to mind. If dopamine fuels an almost addictive love of exploration and physical risk-taking, then what on earth has happened to it? Some 30 per cent of the American population is now obese and appears to have all but lost this drive, preferring inert consumption to effortful consumption. If dopamine has commanded such a powerful drive over our evolutionary history, launching us across the oceans and out into space, why is it now so easily dimmed? We do not yet have an answer to this question, although it remains one of the most pressing issues for medical science. But an intriguing suggestion can be found in some almost-forgotten research conducted in Vancouver in the 1970s, research that came to be known as 'Rat Park'. The seventies were a time when many of the laws governing drugs of abuse were introduced, and the lead author of the study, Bruce Alexander, questioned the logic behind the then-current understanding of addiction.

What he did in his study was place rats in a bare cage with two bottles they could drink from, one of which contained water, the other water laced with morphine. Not surprisingly, the rats preferred the bottle laced with morphine, and in time they became addicted to it. What Alexander did next was interesting. He repeated the experiment, only this time he placed the rats inside what he called Rat Park, a cage with a running wheel, foliage, other rats, both male and female, and so on. In other words, he provided the rats with an enriched environment. When placed in Rat Park the rats did not prefer the morphine-laced water, and did not develop an addiction. In light of later research we could speculate that these rats were getting the daily dopamine hit they needed just from normal forms of search, work and play. Indeed, recent research has found that an enriched environment is such an attractive alternative to drugs that animals addicted to cocaine, once returned to an enriched environment, will actually kick their habit.

Rat Park provides a novel perspective on our current problems of addiction and obesity. It leads us to wonder, have we denied broad swathes of the population an enriched environment? Have we denied them access to sports facilities? To artistic, even scientific, training? To green spaces? Has something gone wrong in the workplace? With urban development? Have we in effect removed large numbers of people from Human Park, and placed them in an empty cage? These are burning questions, because the obesity epidemic, besides being a medical catastrophe, may be dimming the gut feelings and entrepreneurial drive upon which our prosperity and happiness depend.

THE ANNOUNCEMENT

By 1.30 p.m. the market has recovered half its lost ground, with the ten-year Treasury now down only three quarters of a point. Gwen has made back all the money she had lost on her five-year position, and she and Martin have emerged, sure enough, with a tidy profit, amounting to almost $3 million, on the long positions they built up when the market hit its low point. Gwen has her mojo back. No way is she going to be transferred anywhere. (Martin has been reassuring her, saying that Ash is fixated on problems on the mortgage desk, not her.) Gwen and Martin decide to sell out half their positions, reasoning quite rightly that it is not wise to have a large position, long or short, going into the Fed announcement. The rest of the traders have also done well, and the desk now clicks as an experienced team. A mood grows, a happy contagion, a silent complicity. Communication reduces to the occasional half-sentence and monosyllabic affirmation.

A hint of danger may linger in the market, but most traders welcome it, and a slow burn of excitement and quiet confidence prepares them for the Fed announcement. The rumour had been unexpected, the sell-off a complete surprise, but Martin and Gwen have risen to the challenge and surprised themselves with a good profit. These are precisely the circumstances that trigger a surge of dopamine, and this narcotic drug, now flooding their brains, gives Martin and Gwen an incomparable thrill.

The challenge they face is, however, more than an intellectual puzzle. It is a physical activity that demands skill, quick reactions, and metabolic and cardiovascular resources sufficient to support their efforts. So, as the full import of the rumour sinks in, Martin and Gwen's heart rate and breathing speed up, their blood pressure increases, and crucially, stress hormones flood into their blood. Adrenalin releases glucose from the liver, and cortisol energy stores from liver, muscles and fat cells, so that Martin and Gwen have ample fuel supplies to carry them through the afternoon. The cortisol is especially potent, for it freely enters the brain, finds receptors all along the pleasure pathways, and magnifies the effects of dopamine. Physical stressors, like a fast drive, off-piste skiing, or trading an exciting market, provide a thrill not usually expected from a stressor. But at low levels, cortisol, in combination with dopamine, provides a narcotic hit that is nicely described by the neuroscientist Robert Sapolsky as intense stimulation – 'You feel focused, alert, alive, motivated, anticipatory' – and by Greg Berns as a profound feeling of satisfaction.

Indeed, if the demands are high, the outcomes uncertain and the potential rewards substantial, people rise to the challenge with heightened energy and excitement, and with a focused and all-consuming attention that blots out distractions and loses track of time, they enter a state often described by psychologists as flow. People fortunate enough to experience the euphoric state of flow – artists, athletes, mathematicians, others who merely love their jobs – come to live for these moments of heightened performance. With every system of their body switched on and working to perfection, Martin and Gwen feel alive as never before. They revel in their powers and glory in their expertise, life as intense as it gets. This is what they love, this job, this New York, this moment in March.

At 1.45 Martin picks up the mic of the squawk box and provides a commentary for the sales force. His views are widely respected on the Street, and his commentaries give salespeople a pretence for calling clients and drumming up business.

'OK, listen up!' Martin begins. 'We've all heard the rumours and seen the sell-off. We saw massive selling on the way down, but also a hell of a lot of buying down at the bottom, and from clients with very deep pockets. If I had to guess – and that's what I'm paid to do – I'd say a lot of people along the Street got caught short by the bounce and still need to buy bonds. Barring a three-quarter-point rate hike, and no way in hell the Fed's going to do that, I'd trade this market from the long side. Shen will give you our view on the Fed.'

Shen is the bank's resident economist, and he proceeds to give the bank's official view, that they have always believed the Fed would not raise rates today. Having said that, they, like pretty well everyone else on the Street, may have missed one or two hints the Fed has dropped. Shen cannot therefore discount the possibility that the Fed will raise rates today, given its strongly stated disapproval of the developing bubble in stocks, but he is confident that it would not hike any more than half a point.

By 2.10 trading on the screens dwindles. The floor goes quiet. Over the past two hours banks and clients alike have jostled for position, and most have now set their trades. All around the world, traders and investors wait for the announcement. An expectant hush descends on global markets.

I must now try to explain what happens in the next few moments. At 2.14, Martin and Gwen lean into their screens, gaze steady, pupils dilated, their breathing rhythmic and deep, muscles coiled, body and brain fused for the impending action. When the announcement comes across the news wire they will, like tennis players returning serve, spring into action, buying or selling bonds on the screens, volleying client trades, making sense out of the unpredictable market moves; and their activity will most likely continue throughout the afternoon, and maybe into the Tokyo morning. They will thus need to sustain a full fight-or-flight response. But that is not what is happening right now, just moments before the announcement. What happens now is another reflex action, known as the orienting response. The orienting response is an involuntary wait-and-see reaction, and while in its grip our heart and lungs slow almost to a stop.

There is no clear agreement among physiologists on why exactly this happens. When a gazelle stands motionless in the tall grass, hoping not to be noticed by a lion ambling by, its heart and lungs are similarly motionless. Should it be discovered, it instantly leaps into a sprint. But how does it do this? How do its heart and lungs rev up so quickly? It can take a few seconds for the fight-or-flight response to reach its maximum effect, so would it not make more sense for the gazelle, while waiting, to crank its heart and lungs up to full capacity before the lion notices it? Then when it needs to run there would be no delays.

, One explanation for the orienting response is that the heart and lungs slow in order to increase the size of their draw, so that when we do jump into action the lungs are full of air and the heart full of blood. This is undoubtedly true. But there may be another reason. When a gazelle stands motionless watching a nearby lion, when a sprinter crouches at the starting line, when a goalkeeper prepares to defend a penalty kick, and when Martin and Gwen sit motionless before the Fed announcement, their bodies are indeed preparing the fight-or-flight response. They are, all of them, geared up and ready to go. Just not yet. For their heart and lungs are being held back, like an attack dog straining on a leash, by their vagus nerve.

During the orienting response both the fight-or-flight and the rest-and-digest nervous systems are activated at the same time. When this happens our body is fully prepared for a quick sprint or a fight to the finish, but it is being restrained by what Stephen Porges, of the University of Illinois at Chicago, calls the vagal brake. At this point your body resembles a drag-racing car at the starting line: it locks its front brakes but guns its engine and spins its back tires, burning rubber and shooting out flames, until at the green light the driver merely releases the brake, and the car rockets down the track. The principle here is that the car accelerates far faster by releasing the brake on an already revving engine than it does by stepping on the gas and initiating acceleration. Much the same thing seems to happen in your body. The vagus, a powerful and fast-acting nerve, restrains the fight-or-flight response, allowing blood pressure and circulating levels of adrenalin to build up, and then releases its brake when the news comes out, instantaneously bringing heart and lungs to full speed.

2.15 p.m. A single line prints out across the news wire. A brief pause, then Shen yells over the squawk, 'A quarter point! The Fed raises rates a quarter of a per cent.' The floor takes a moment to digest the news.

Another line of print follows, containing the text of the Fed's announcement and its reasons for acting, saying in effect that it remains vigilant about excesses in the stock market. Then the vacuum created by the momentary confusion fills with an explosion of activity. Vagus nerves across the globe release their brakes, and everyone starts screaming at once. The screens flicker with prices, and the squawk jumps to life with trades flooding in from clients, some selling, others buying, some of them the very clients who were desperate to sell this morning. There is utter confusion. The Fed has indeed raised rates, when few people were expecting it to, and that brute fact is the first to register. The market trades down, with the ten-year dropping an instantaneous half-point, but without much volume and not a lot of trades taking place. Merely a kneejerk reaction.

But then the more considered response asserts itself – a quarter point? That's nothing. It's not three quarters of a point, nor even half a point. What's all this tough talk about reining in the stock market? A quarter point? That's nothing! A slap on the wrist. This Fed is not putting up a fight; it is not going to back up its tough talk with aggressive rate hikes, not like Paul Volker did back in the early eighties when he hiked rates to over 20 per cent. Now that was one tough banker. But a quarter point? Ha!

The market breathes a collective sigh of relief, and then it is off to the races. Stocks, emboldened by the news, rally a quick 200 points, and the bond market regains all its losses for the day and then continues to rally. The trades now coming in through the sales desks are large, non-stop, and all on the buy side. Martin and Gwen use these requests for bonds as a chance to scale out of their remaining long position, bought almost two points below the current market price of 101.16 on the ten-year. The fun continues all afternoon, and towards the end of the day Ash strolls by and whispers to Martin and Gwen that the floor is on track for a great day. The Treasury desk, together with its satellite desks such as Treasury bills and government agencies, looks as if it will haul in almost $12 million.

The Fed has now lost its power to scare. In its defence it has to be said that this loss of credibility is not entirely the Fed's fault. Central banks face a near-impossible task when confronted by irrational exuberance, for at times like these controlling the market is like herding schoolchildren on a sugar high. Besides, if it raises rates enough to stop a stock market bubble, it could just as easily kill the economy. This is a very real risk, for when a market is aflame and maddening for profits, raising rates a percentage point or two may have very little effect on the market, but would have an enormous impact on other businesses that depend on borrowed money, like manufacturing and utilities. If investors think the latest tech stocks, for instance, will prove to be the next IBM or Microsoft, they are not going to be dissuaded by a couple of per cent of extra interest they could earn on bank deposits. And that is precisely the mentality that has taken root in the markets.

The seeds have been sown for a good bonus season. At the end of the day traders on a high float off the trading floor and head out to celebrate their triumphs with martinis, Nouveau Mexique and a late night in clubs throughout SoHo and Tribeca, neighbourhoods where the myth of Bohemia lives on even though the arts have long since been killed off by the Midas touch of Wall Street. Logan, hoisting his gym bag, announces confidently to a couple of salespeople lingering at their desks that nothing stands in the way of the Dow reaching 36,000.

SIX

The Fuel of Exuberance

As March slips away and winter's chill lifts, a spirit of youthfulness descends on Wall Street. With the Fed's voice of authority all but silenced, the markets take on the appearance of an unsupervised playground. A bull market over the past two years may have taken stocks up 40 per cent and bonds about 20 per cent, but traders and investors believe this is just the beginning. With wild-eyed enthusiasm they conclude that a historic epoch, amounting to nothing less than a renaissance for the American economy, with permanently high growth rates and low inflation, is dawning, so bonds and stocks caper from high to high. News, no matter what its content, arrives with the breathless promise of unparalleled opportunity. Financial journalists warn investors against dithering, and declare that now is seed time in the fields of investment.

When markets are in this giddy state, money rains down upon Wall Street. Every department of an investment bank, no matter if it is directly involved in the rallying markets or not, starts to report record profits. The securities that banks underwrite – the stocks, the corporate and mortgage-backed bonds – inevitably rally in price, so the banks make a fortune on any unsold inventories. And clients, the real money behind the markets, sit on assets that have increased dramatically in value, feel successful, so they become less aggressive in the prices they demand from Wall Street, leaving money scattered on the table. The extra margins made on client transactions can add up to a record year and massive bonuses for the bankers.

With these outsized profits comes an irresistible urge among traders to increase the amount of risk they take. Where once a trader may have been comfortable trading $100 million bonds, he now trades $200 million, even $1 billion. And with these increased position sizes, coupled with strongly rallying markets, come unexpectedly large trading results, known in banks as the traders' P&L, short for profit and loss statement. Where once a trader may have averaged a P&L of $250,000 per day, he now posts $375,000. The difference, should it be maintained over the 230-odd trading days in the average banker's year, adds up to an extra $29 million in P&L, and maybe an extra $3 million in bonus.

This upward drift in risk has occurred in most bull markets, but during one in particular it assumed well-nigh incredible proportions. During the recent housing bubble, roughly between 2002 and 2006, the financial world experienced nothing short of a hyperinflation. Before that time, during the 1990s, a golden boy on Wall Street might have traded a risk-weighted equivalent up to $500 million ten-years, attained P&Ls in the $30–50 million range, and been paid bonuses in the $1–3 million range, the real stars up to $5 million. Yet during the noughties, only a few years later, it was as if the financial world had added a decimal point to every number it dealt with, to position sizes, P&Ls and bonuses, which could now amount to over $50 million. I had been quite successful as a trader, and was accustomed to taking big risks, executing what at the time were some of the largest Eurodollar options trades on the Chicago Mercantile Exchange, but when I visited my old haunts in 2005 I barely recognised this new world. I felt like a grandfather recounting war stories from the days of cavalry charges, and receiving patronising smiles in return. But most sensible observers thought the risks being taken during the housing bubble were dangerous and ill-conceived. Tragically, they were right.

While they last, though, bubbles like these can be fun. Unexpectedly large P&L emerging from greater than usual risk-taking is precisely the kind of situation that causes dopamine to surge into the nucleus accumbens, this narcotic most likely providing traders with the euphoric hit they enjoy during bull markets. But as the rally progresses, traders feel something else added to the mix, something more profound, more physical, like the rumble of some large engine switching on. For as profits rise, so too does testosterone. In fact, the two systems, the dopamine and the testosterone, are thought to work synergistically, with testosterone achieving its exciting effects largely by increasing dopamine in the nucleus accumbens, testosterone constituting the bass line of euphoria, dopamine the treble. Indeed, some evidence suggests that sex steroids such as testosterone may sensitise the brain to the effects of dopamine, making all rewards, from winning in sport, to victory in battle, to a large P&L, that much more thrilling; further evidence even suggests that steroids can be addictive.

Testosterone, being a steroid hormone, works on a slower time scale than most molecules we have looked at. Unlike adrenalin, for example, which is pre-produced and stored in little pouches called vesicles, waiting to be released, steroids cannot be stored. Steroids are molecules that can cross cell membranes, even permeate skin (many steroids, such as testosterone, are applied as a gel) or penetrate the rubber gloves of lab technicians. Trying to hoard steroid molecules in a vesicle would be like trying to lock ghosts in a room – they would just drift through the cell walls. As a result they are produced only when needed, and then released into the blood, a time-consuming process. So the hormone signalling process, from hypothalamus to the production of steroid hormone, takes up to fifteen minutes just to get started.

It takes even longer for steroids to take effect – hours, even days. The process may be slow, but the way steroids work is unique in the human body. They cross membranes, enter the cell nucleus, and cause gene transcription. In other words, steroids cause proteins, the building blocks of the body, to be manufactured. Furthermore, unlike other hormones which generally have effects localised to one or two tissues, steroids have receptors in almost every nucleated cell in the body. All these properties of steroids give you an inkling of their power. A single steroid like testosterone can cause a bewildering suite of physiological changes, building up bone density and lean-muscle mass, increasing haemoglobin and clotting agents in your blood, heightening mood, tormenting you with sexual fantasy, and tilting behaviour towards greater risk-taking. By doing so, testosterone orchestrates a focused and coordinated physical response to the competition and opportunity at hand. So during a bull market traders feel a deeper, more far-reaching transformation as the big engines of their physiology start to turbocharge their risk-taking.

HOW MEN ARE MADE

It can be difficult to view testosterone as a molecule, and a subject fit for serious scientific and medical research, because it comes shrouded in myth and cliché. The mere mention of the word seems enough to dispel any air of scientific objectivity. Since the early days of research into sex hormones the findings have too easily slipped out of the hands of scientists and clinicians and into the hands of quacks, who claimed they had bottled the fountain of youth, the ultimate aphrodisiac, the magic potion, like that possessed by the Gauls in the *Asterix* comic books, which promises superhuman powers on the field of battle. Unfortunately, many of the scientists who discovered this molecule contributed to the fog of hype shrouding it.

In 1889 a French-American neurologist by the name of Charles Edward Brown-Séquard mixed up a witches' brew from the testicles of dogs and guinea pigs. Brown-Séquard was a well-respected medic, and to this day a disorder of the nervous system he identified bears his name; but when this 72-year-old scientist downed his own concoction of animal testicles he lost all scientific impartiality, claiming to have found a 'rejuvenating elixir' and proudly confiding to an audience in Paris that just that day he had 'paid a visit' to his wife. It is now thought that the virility of which he boasted was largely the result of a placebo effect, but his pronouncements stamped research into hormones in general, and testosterone in particular, from that day forth with crazed and extravagant expectations. Later, in the 1920s and 30s, the search for the active ingredient in animal testicles turned into something of an arms race, and was joined by scientists around the world. Research into hormones at this time seemed to hold out the promise of a superhuman life through chemistry. So hot was this research that it showed up in magazines and popular culture. Noël Coward, the master of the drawing-room comedy, picked up on this chatter; and in his 1932 play *Design for Living* has one character, Ernest, say, 'I wish you'd tell me what's upsetting you,' to which he receives the reply, 'Glands, I expect. Everything's glandular. I read a book about it the other day.'

It was in the previous year, 1931, that the first androgen – the class of steroid to which testosterone belongs – was isolated. A German scientist named Adolf Butenandt managed to extract 50 milligrams of androsterone, a weak form of testosterone, from 25,000 litres of urine donated by a police barracks in Berlin. He and others believed this chemical had important medical and commercial applications, but there had to be a better way of manufacturing it. Pharmaceutical companies devoted a great deal of time and money trying to synthesise the hormone from its mother molecule, cholesterol, something Butenandt and a Croatian scientist named Leopold Ružička succeeded in doing in 1935. For their efforts they were jointly awarded the Nobel Prize for chemistry in 1939, achieving the acme of scientific respectability. Yet the scientific breakthrough continued to be accompanied by breathless hype: when describing testosterone to the Nobel Committee, Butenandt exclaimed, 'Dynamite, gentlemen, it is pure dynamite!'

By the end of the 1930s this molecule was already at work in medical clinics, being used to treat depression and what was then called 'involutional melancholia', the ebbing in vitality among men entering middle age, often the result of a perfectly natural decline in testosterone. Today this condition – although if it is natural ageing it should perhaps not be called a condition at all – is marketed by pharmaceutical companies as andropause, the male equivalent to menopause, although the term has yet to achieve medical respectability.

* * *

So, all hype aside, what exactly is testosterone? Testosterone is commonly thought of as a male sex hormone, but it is found in women as well. There are profound differences, however, between the sexes. Men produce it in their testes and to a lesser extent their adrenal glands, and women in their ovaries and adrenals. Importantly, men produce about ten times the amount of testosterone as women, and therefore display more pronounced effects. In fact, so broad and powerful are these effects that it is this molecule, almost on its own, which creates a male. Let me explain.

We each have twenty-three pairs of chromosomes, and it is the twenty-third pair which determines whether a foetus is a boy or girl. This chromosome can be either XX, in which case the foetus develops as a female, or XY, in which case it develops as a male. The default sex of all foetuses is female: unless they have a Y chromosome, they will develop into females. The Y chromosome is a surprisingly simple one, with very few genes on it. One of these is responsible for the bulk of the differences between men and women. This gene is called SRY, standing for Sex Determining Region of the Y chromosome.

What the SRY gene does is simple. It codes for the building of a protein hormone called Testis-Determining Factor, which shunts the primordial gonads off the path leading them to develop into ovaries and onto a path leading them to develop into testes. Once they start to grow, the testes produce testosterone, and this molecule creeps out into the bloodstream and does all the rest of the work, docking in receptors throughout the body and morphing tissues into a male rather than female form. And that is about it. One gene, one protein hormone, growth of testes, then testosterone does almost all the rest, creating a male from Eve's rib. Recently scientists have found other genes that code for differences between men and women, especially their brains, but testosterone nonetheless does most of the work. That is one potent chemical, just as Brown-Séquard and Butenandt advertised.

Problems lurk in this Y chromosome. Chromosomes normally swap genetic material, a process known as recombination, and this exchange has the felicitous effect of repairing any damaged genetic material, ensuring our continued health. Genetic recombination can be compared to the regular servicing you schedule for your car, in which old parts are replaced by new ones. Our chromosomes do much the same thing when they recombine – they exchange old and broken genetic parts for new ones. An X chromosome can swap material with another X chromosome, thus ensuring that each generation is fitted with new parts. But not so the isolated Y. This lone wolf has nothing it can swap with, so over time, like a car that is never serviced, it compounds problems and accumulates damage until its genes, one by one, die off. Some animals, such as the kangaroo, now have only a few genes remaining on their Y chromosome. This slow death of the Y has been called Adam's Curse by the Oxford geneticist Bryan Sykes, who predicts that in 5,000 generations men will be extinct.

Testosterone levels fluctuate dramatically over the course of a man's life. There is a prenatal surge between the eighth and nineteenth week of gestation, and it is at this time that the foetus is masculinised, the testosterone spreading out across the body and brain and creating the tissues, chemical circuits and receptor fields that will influence adult male behaviour. Testosterone levels then subside, spike once more just after birth, for reasons that are not well understood, and then subside again until puberty, this hormonal vacation allowing little boys to be the angels that they are. At puberty testosterone comes flooding back into a boy's body, coursing along canals, activating tissues that the testosterone itself had created and then left for years to lie dormant, much like sleeper cells in a spy ring. Its effects now are profound, building muscle, manufacturing sperm, lowering the voice, growing facial hair, stimulating sebaceous glands in the skin and frequently causing acne. Later, beginning in his early thirties, a man's testosterone levels begin to fall, and continue to do so over the rest of his life, these falling hormone levels perhaps calibrating the risks he chooses to take with his body's declining ability to handle them.

The surge of testosterone at puberty can drive much of the risky behaviour teenage boys are known for. Blame cannot, however, be placed entirely on hormones, because the teenage brain has not finished developing, and there is some evidence that the nucleus accumbens, the thrill centre of the brain, outgrows the more rational prefrontal cortex well into our twenties. Whatever the cause, teenage boys are a menace, and every adult male, myself included, knows deep down that he is lucky to have survived those years.

There is another noticeable effect of these testosterone surges, and that is on sex drive. In animals this hormone prepares a male both to fight and to mate during breeding season, this dual action illustrating yet again how steroids unify body and brain during archetypal moments of life. Among humans testosterone has much the same (albeit muted) effect, on both men and women – it increases desire and sexual fantasy. It should be noted, however, that while testosterone affects a man's desire, it is not directly involved in the mechanics of erection. Erections are controlled, oddly, by the rest-and-digest nervous system (that is why having sex can be difficult if you are stressed), while ejaculation is controlled by the fight-or-flight nervous system. Sex thus requires a complicated synchronising of hormones and two branches of the nervous system. Testosterone, though, mainly affects a male's interest in sex, his tendency to think about it every few minutes, to find sexual cues everywhere, and to spin out maddening fantasies.

What determines the amount of testosterone a foetus is exposed to in the womb? These levels are largely the result of genetics, as one would expect, but there is some evidence that birth order also affects them. In animals, the first-born in a clutch of offspring has a distinct advantage over its siblings, a day or two head-start making it larger and stronger, and capable of hogging the food and even pushing the next-born out of the nest. Nature, however, has found a way of evening the odds of survival. A mother's body, in the words of one research team, 'appears to "remember" previously carried sons', possibly because each of them leaves behind a marker, known as the HY antigen. A female, a bird for example, will then deposit higher levels of testosterone in later-born males. These males may be smaller at birth, but they are meaner, and that evens the odds with their bigger siblings. Some reports suggest that the same mechanism exists in humans, with younger sons often proving more aggressive than their older brothers.

What physical advantages do these later-born males enjoy? Developmental biologists distinguish between the anabolic and the masculinising effects of testosterone. Masculinising effects include the growth of facial hair, the lowering of the voice, and the growth of testicles and sperm-producing cells. The anabolic effects include an increase in lean-muscle mass, in haemoglobin, and in bone density. It is the anabolic effects that athletes are after when they illicitly take steroids.

Today vast sums of money are poured into the design and manufacture of synthetic androgens. The result is an extensive menu of anabolic hormones known in the gym as 'roids', 'juice', 'hype', 'pumpers', 'gym candy' or 'Arnolds'. A fascinating glimpse into this world of illicit steroid use can be found in the film *The Wrestler*, in which Mickey Rourke plays an ageing fighter who relies on drugs to keep up his form and strength. Athletes abusing anabolic steroids, frequently raising their levels of testosterone by four to five times those naturally found in their bodies, can suffer some unexpected side-effects. For example, their testicles, interpreting the excessive levels of testosterone in the blood as a signal that they do not need to produce any more, shrivel to the size of raisins. Furthermore, in many tissues of the body and brain, testosterone, through a curious twist of evolutionary fate, must be transformed by an enzyme called aromatase into the female sex hormone oestrogen before it can become biologically active. Adipose tissue – fat cells really – are particularly rich in this enzyme, so men who have either a lot of these cells, such as those who are obese, or a lot of testosterone, such as those using anabolic steroids, can end up growing breasts, a condition known as gynaecomastia. At sufficiently high levels, testosterone, the ultimate male molecule, bizarrely starts to feminise weightlifters.

At moderate levels, though, anabolic steroids give an athlete a very clear advantage, and this has led to a high-tech cat-and-mouse game between sports scientists trying to elevate testosterone levels illegally and the policemen of governing bodies like the IOC. The clear advantages testosterone bestows on athletes have also led to a much more controversial form of testing, the testing for an athlete's sex. It had

157

been suspected in some Olympic Games that gold medals in some women's events had actually been awarded to men, so the practice began of testing female athletes to see if they were indeed what they claimed to be. At first this seemed a simple case of looking for a Y chromosome. If an athlete tested positive for XY, then he was a male. Simple. As a result of this test a handful of women at both the 1992 Barcelona and the 1996 Atlanta Olympic Games were disqualified.

Unfortunately, the reasoning behind the tests was questionable. If a person has a Y chromosome then they do in fact produce testosterone, starting at the foetal stage. But what if that person has been born with a genetic disorder that makes them insensitive to testosterone? What if their testosterone receptors do not work? Then the testosterone will have no effect. This is precisely what happens in a condition called Androgen Insensitivity Syndrome – people with a Y chromosome produce testosterone, yet it does not masculinise them. Since the default sex of a foetus is female, these people will appear to all the world as women. Are they women? They think so, and who is to tell them otherwise? In the end the issue of establishing sexual identity became intractable and politically fraught. Many women athletes found the tests invasive, far too public, and almost medieval in their humiliation, like displaying the bedsheets of newlyweds. As a result, sex testing was dropped from the 2000 Sydney Olympics.

TAKING THE HEDGE OUT OF HEDGED TRADES

This is a glorious time for the trading desks: the flows are deep, the markets volatile yet optimistic. The Treasury traders find themselves taking larger and larger positions and making more money than ever before. But the risks they take are as nothing compared to those taken on other desks. The real risks to a bank's capital, and to its solvency, are usually found on the desks trading securities that carry credit risk, securities such as stocks, corporate bonds (ones issued by private companies), junk bonds (ones issued by private companies that are on the verge of bankruptcy), and mortgage-backed bonds.

There is one desk, though, that has a mandate to trade all of these dangerous securities, and not surprisingly it often lies at ground zero of any financial crisis. This desk is located just down the aisle from Martin – the fixed income arb desk. Arb is short for arbitrage, a complicated and supposedly clever type of trading designed to profit from incorrectly priced securities. Arb traders do not provide prices or services for clients, as do flow traders like Martin and Gwen; rather, they trade for the bank's own account, buying securities that seem cheap and selling ones that seem expensive, sometimes amassing huge leveraged positions, that is, ones financed with borrowed money. The positions can end up being several times larger than the entire value of the bank itself. To put this leverage in perspective, these traders' positions can be compared to a home-owner who borrows $20 million against the collateral of his $500,000 home, in order to buy some rental properties. Were the rental properties to drop in value by a mere 2.5 per cent, the home-owner's capital would be wiped out, bankrupting him. It was leverage on this scale that in 2008 bankrupted the investment bank Lehman Brothers.

Arb desks house the so-called rocket scientists, ex-physics or engineering quants who build models designed to spot pricing anomalies in, say, the yield curve or the volatility surface in the options market. Stefan, for example, the head of the desk, has a Ph.D in physics from the University of Moscow and has worked on Superstring Theory, a bewildering branch of quantum physics in which sub-atomic objects vibrate in ten dimensions. Arb traders, being inscrutable geniuses, have a licence to act inappropriately – to go unshaven, dress down, show up for work at noon. Unlike flow traders, who must be clean-cut, well-dressed (but not too flashy – Brooks Brothers rather than Prada) and capable of observing basic table manners when entertaining clients, arb traders are nurtured as the floor's eccentrics. Whether or not they have the brains and personalities to carry this persona does not matter. Trading floors have a fixed mythology – the funny guy, the sports legend, the brainy eccentric – and people get slotted into one category or other, whether they belong there or not. So it is that arb traders get typecast, tolerated and, tragically, given the benefit of the doubt when management does not understand what they are doing.

One of the arb traders, Scott, has a particularly valuable expertise: spotting discrepancies between the value of a company's stock and its bonds. The stock of GM or GE or IBM may reflect optimistic expectations of future earnings, while its bonds may reflect worries about the company's finances. In these cases the stock may be priced too high and the bonds too low, presenting Scott with an arbitrage opportunity. Scott is particularly skilled at this sort of trading because he is a former flow trader from the corporate bond desk, and has extensive experience in assessing credit quality. As a result he is one of the few people in the bond department permitted to trade actively in the stock market. When Scott spots a situation in which two securities are mispriced relative to each other he buys one and shorts the other, establishing what is called a spread or a hedged trade – the terms are mostly synonymous – one of the most common trading strategies pursued by arbitrage desks and hedge funds. They are also used extensively by flow traders such as Martin.

What exactly is a spread trade? What is hedging? What is shorting? Since few people outside banking really understand these trading strategies, it may be worthwhile to explain them. They are also well worth understanding because many of the most toxic positions taken by financial institutions, the positions that frequently lead to crises, are these spread or hedged trades. This fact carries a certain irony: a spread or hedged trade is one that is supposed to minimise market risk. Because a spread trade profits from the mispricing of two securities relative to each other, its profits do not in principle depend on the market going up or down. They depend rather on the price difference between the two securities returning to normal, and this has always been assumed to be a less risky type of trade. To see how a spread trade works, consider an example set in your local fruit market.

Imagine a situation in which an orange is worth on average 10¢ more than an apple. If oranges at the market are priced at, say, 60¢ each, then apples would normally be priced at 50¢. Imagine as well that this 10¢ price difference persists even when the price of fruit goes up or down: it might rise in value with inflation, with oranges increasing to $1, but then apples would also increase, to 90¢. This price relationship is reliable enough to assure you that if the prices of the two fruits diverge from the 10¢ difference to, say, 20¢, then you know with some assurance that eventually they will return to the normal 10¢ difference.

Now imagine you are strolling through the market one day and you notice that oranges are priced at 60¢, yet apples are at 40¢, a difference of 20¢. You fully expect this price difference to revert to normal; and being crafty by nature, you try to figure out how to make money from that prediction. You could buy apples at what seems like a cheap price of 40¢; but if fruit were to fall in value, because of, say, a bumper harvest, then apples could drop to 20¢. In this case, even if the price difference between apples and oranges reverted to 10¢, with oranges dropping to 30¢, you would have lost 20¢ on your apples. Similarly, you could sell oranges at the seemingly high price of 60¢, but if the fruit market were to rally in price, perhaps because of a frost in the fruit-growing regions, then oranges could rise to $1, and apples to 90¢. Here again the price difference has reverted to normal, but you have lost money on your oranges, in this case 40¢. The only way to profit from the price discrepancy returning to normal is by establishing a spread or a hedged position, one whose profits are determined independently of the market going up or down. A hedged position is therefore said to be market neutral.

How does it work? How could you, as a trader, profit from the mispricing of fruit in a market-neutral manner? You could do it with three separate transactions: First, you buy, say, 100 apples at 40¢ from one fruit stall. Second, you sell 100 oranges at 60¢ to another stall which is looking to buy them. Selling an asset you do not own is called shorting. But how do you deliver 100 oranges to this stall if you do not own them? Selling something you do not own constitutes fraud, so you have to deliver the oranges somehow. To avoid breaking the law you undertake the third transaction: you go to another stall, take its owner to one side and confide in him. 'Look,' you say, 'I need to borrow 100 oranges for a few days. I don't want to buy them, just borrow them. If you lend me 100 oranges I'll replace them in a few days, and for your services I'll also then pay you 5¢ per orange.' This stall owner, looking at his stock of oranges, thinks the proposal is a reasonable way of making a bit of extra money on fruit that might otherwise just sit there. So he lends you the 100 oranges, which you in turn deliver to your buyer.

You have now put on a spread trade. And look at the finances of your small hedge fund. You bought 100 apples at 40¢, so you pay out $40; but you also sold 100 oranges for 60¢, so receive $60, and you can use that to pay for the apples. The arbitrage position has been put on with no capital required (although in the financial markets you are required to post a small amount of margin or collateral), meaning that you could in principle amass enormous positions and eventually dominate your local fruit market. This simple game of leverage permitted the hedge fund Long Term Capital to leverage their capital into such large positions that when they went wrong they wiped out the company's capital and threatened to destroy the entire financial system.

Your trade is tiny in comparison to those of even the smallest hedge fund. Nonetheless, over the next few days the fruit market goes on a rollercoaster ride: rumours one day of a good harvest cause prices to drop 25 per cent; rumours the next day of a strike among migrant fruit pickers cause them to spike 50 per cent. No matter – apples and oranges go up and down together, so what you lose on one leg, as it is called, of your hedged position you make on the other. At one point, however, the price spread did exceed 20¢, widening to 25¢, with apples going down in price and oranges going up, and this gave you the jitters. There is no scarier predicament in the financial markets than a spread trade in which the two securities become uncorrelated and start moving in opposite directions, something that unfortunately tends to happen during crises. Hedged positions are billed as market neutral, but the term is something of a misnomer, because they carry substantial risks.

But this week in the fruit market the correlations hold strong, and after a few days the old price relationship of a 10¢ difference reasserts itself, with apples at 30¢ and oranges at 40¢. Now you decide to unwind your trade. You sell the 100 apples you bought at 40¢ at the new price of 30¢, losing 10¢, for a total loss of $10; you buy the 100 oranges you sold at 60¢ at the new price of 40¢, making 20¢, for a total profit of $20; you give back the 100 oranges you borrowed and pay the stall owner $5. Summing up, you have made $5 on your spread trade, for little risk (or so the thinking goes) and minimal use of capital.

If you understand this simple example of shorting and hedging and spread trades, then you have in your hands the basic principles of high finance. Spread trades are the staple trading strategy of both banks and hedge funds. Within the banks it is not just the arb desk that employs them, for flow traders do too, on the Treasury, mortgage, corporate bond desks. Flow traders facilitate customer trades, but they often use these customer flows as a means of building up large spread trades. For example, if Martin had believed that the ten-year Treasuries he sold to DuPont were expensive relative to, say, two-year Treasuries or bond futures, he could have bought either of these securities instead and established a large spread trade. Flow trading desks were originally intended mainly to serve clients, but over time the arbitrage positions they held grew exponentially in size and P&L, until during the housing bubble years they dwarfed the client trades.

To make the jump from the fruit market into high finance, however, you have to stretch your imagination to consider spread trades between, say, Treasury bonds and mortgage-backed bonds, or between Treasuries and stocks, or between German and Greek bonds, or gold and silver, even the amount of rainfall in California relative to Kansas. In addition, you have to consider numbers that verge on the incredible – aggregate positions running into the trillions and profits into the hundred millions, even billions. Your small spread trade made $5, but if you were to scale up your position to a size comparable to the big hedge funds, and bought, say, $4 billion-worth of apples, then your profit would have amounted to $500 million, a decent week in the fruit market.

*　*　*

163

And this is precisely what Scott has been doing this year, setting up spread trades, mostly between the stock and bond markets. He believes strongly that stocks are undervalued, so he has been buying them, and that bonds, especially corporates, are overvalued, so he has been shorting them. He has met with unprecedented success at this form of arbitrage, and can boast a P&L year to date of almost $17 million, an exceptionally strong showing given that it is early in the year. Scott does the mental arithmetic, calculating his year-end P&L if he keeps making money at this rate – maybe $40 million – and warms to the idea of a $5 million bonus at year end.

But Scott's success has led him to throw caution, the most valuable trait in an arb trader, to the wind. He is now convinced that all his spread trades are in essence nothing more than a cumbersome way of buying the stock market. He is one of the many traders who has taken the Fed's message as a green light for a market rally, so recently, in addition to the spread trades he maintains as his core strategy, he has spent more and more time doing nothing more sophisticated than buying stocks and waiting for them to go up. So far he has displayed an almost occult ability in calling market direction, so his boss Stefan has let him continue.

Specifically, Scott bought futures contracts on the S&P 500 index, contracts called e-minis. His initial position, put on after the Fed announcement, amounted to 2,000 contracts, meaning that for a 1 per cent move in the stock market he makes or loses about $1.3 million, a reasonable-sized trade, but nothing too dangerous. Scott's strategy has worked well: the stock market has rallied almost 100 points, making him $10 million.

Cooler minds would conclude that the prudent thing to do now is unwind his core position, take some profit off the table, or at the very least scale down its size. But that does not seem to happen during bull markets. Scott is certainly not thinking that way. Far from it. He can barely contain his excitement. For Scott has seen hove into view a rare trade, the big one, his great white whale. Those who deal in the lore of the market refer to this opportunity as 'the retirement trade', the one that will make so much money they will never have to work again.

The retirement trade, it should be noted, is something of a myth; not because such opportunities do not exist – they do – but because very few bankers, despite their professed dream of spending the rest of their lives on a golf course, actually want to leave this game. They would miss the excitement, the feeling that they are at the centre of the world, which in a sense they are. Most would find it difficult to replicate the dopamine rush they get from trading.

Stocks may have rallied a quick 15 per cent this spring, taking their dividend yield down below 2 per cent, but, Scott argues, just look at the macro-economic backdrop! Asia has only just started to grow, as have Africa, the Middle East, Latin America and Eastern Europe (the Russian default of 1998 quickly fading from memory). At no time in history have so many geographic regions been brought within the ambit of the world markets. This is George W. Bush's New World Order, and in such a world old valuation tools cannot be applied. Scott believes the market is on the cusp of being revalued.

Something similar happened back in the early 1980s. At that time, the price–earnings ratio of the S&P index amounted to little more than 9 – a low number compared to its long-term average of 15 – largely because many investors had lived through the Crash of 1929 and the Great Depression, and retained a deep-seated fear of stocks. But as this generation aged and lost their clout in the market, dark memories of financial crisis that had haunted the economy for fifty years began to fade. After the 1970s fewer and fewer people remembered the hard times of the thirties, and took to spending and investing with the same blithe spirit as their grandparents in the 1920s. Stocks proceeded to rally for the next 25 years, taking their price–earning ratios up to a peak of 44 in 1999. The aftermath of the Tech Crash of 2001 has, Scott believes, presented a similar opportunity. Doubts and fears linger among survivors of that crash. But not in Scott. He decides against reducing his core position, and instead builds on it.

THE WINNER EFFECT

The euphoria, overconfidence and heightened appetite for risk that grip traders during a bull market may result from a phenomenon known in biology as 'the winner effect'. I first heard of this model during the dot.com years, while listening to a lecture at Rockefeller, and I thought, as I have explained previously, it was the most compelling model of irrational exuberance I had encountered. It was this model that led me back to research, to see if the winner effect exists in the markets.

How does the winner effect work among animals? Does it exist in humans? Can it explain the heightened risk-taking and manic behaviour I had witnessed on Wall Street during the dot.com bubble?

Biologists studying animals in the field had noticed that an animal winning a fight or a competition for turf was more likely to win its next fight. This phenomenon had been observed in a large number of species. Such a finding raised the possibility that the mere act of winning contributes to further wins. But before biologists could draw such a conclusion they had to consider a number of alternative explanations. For example, maybe an animal keeps winning simply because it is physically larger than its rivals. To rule out possibilities such as this, biologists constructed controlled experiments in which they pitted animals that were equally matched in size, or rather that were equally matched in what is called 'resource holding potential', in other words the total physical resources – muscular, metabolic, cardiovascular – an animal can draw on in an all-out fight. They also controlled for motivation, because a small, hungry animal eating a carcass can successfully chase off a larger, well-fed animal. Yet even when animals were evenly matched for size (or resources) and motivation, a pure winner effect nonetheless emerged.

Research on the winner effect began with this statistical finding, but it lacked an explanation. How can winning contribute to further wins? Some scientists argued that a victory imparts information to an animal about its own resources and abilities relative to those of rival animals, and this knowledge permits it to choose fights it can win.

Others thought that winning may leave physical traces, such as phero-mones and other chemicals, on an animal which broadcast its recent victory, and these can deter subsequent opponents from escalating an encounter. But perhaps the most persuasive account is one that high-lights the role of testosterone in these competitions.

When two male animals face off they experience a pronounced rise in their testosterone levels. Testosterone's allotted role in male bodies is to prepare them for precisely these sorts of confrontations. Hence the anabolic effects on muscle mass and haemoglobin. Testosterone has also been found to quicken reactions, to sharpen a type of visual skill known as visuo-motor scanning, and to enhance another visual ability known as camouflage breaking. Just as important as the physi-cal preparation is the hormone's tendency to increase an animal's persistence and fearlessness. After all, there would be little point in equipping an animal with a greatly augmented fighting capacity if it were not willing to use it.

Testosterone thus prepares an animal for a competition, but it is what happens next that drives the winner effect. After the fight is over the winning animal emerges with even higher levels of testosterone, the loser with lower levels. These hormonal signals, it has been argued, make sense: if you have just lost a fight, you had better retire into the bush and nurse your wounds; while if you have won you can expect an increased number of challenges to your newly elevated status in the social hierarchy. Among animals these effects can be large. In one study it was reported that in a competition for rank among recently introduced rhesus monkeys, the winner emerged with a tenfold increase in testosterone, while the loser experienced a drop to one tenth of base levels, and these new levels for both winner and loser persisted for several weeks. So powerful is this effect that in some species dominant males will intervene in a fight, apparently protect-ing the weaker male, but with the intention of depriving the likely winner, and potential future rival, of the benefits of the winner effect.

In many species, male animals display a dazzling ornamentation – bright colours, fancy wattles and combs, lush nuptial plumage – and ostentatious behaviour, strutting and threatening. Some lizards, for example, when preparing for a fight, bob their heads and pump themselves up and down on their forelegs, much like a triumphant athlete punching the air. But the effects of winning and losing can alter these animals' physical appearance. In defeat they can go into rapid decline, their bravado dimmed, some even see their colours wash away, their testicles and brains shrink, and they lapse into a lethargic, even a depressed, state.

Life for the winner is more glorious. It enters the next round of competition with already elevated testosterone levels, and this androgenic priming gives it an edge that increases its chances of winning yet again. Through this process an animal can be drawn into a positive-feedback loop, in which victory leads to raised testosterone levels which in turn leads to further victory.

Does the winner effect occur in humans? The question is controversial. Many social scientists have denied the existence of winning streaks or what are called 'hot hands' in sports, claiming that athletes and fans who believe otherwise are subject to an illusion. But I think the winner effect does exist among humans. First, let us look at the stats. My colleague Lionel Page and I have done just that. To do so we followed as closely as possible the protocol used by biologists when studying animals in the wild. Lionel searched for athletes who were equally matched for physical resources and motivation, and then tested to see if winning on its own contributed to further wins. He came up with what I think is the most rigorous test of the winner effect in sport. Lionel managed to find a database of 623,000 professional tennis matches; from this sample he considered only those between players who were one seed apart; then within this subset he further narrowed his search to matches that went to a tie break in the first set; and even further, to tie breaks decided by the narrowest possible margin, two points. In short, he looked at tennis players who were as closely matched as possible in rank and in playing form on the day of the match. Studying only these close matches, he found that the winner of the first set had a 60 per cent chance of winning the second one, and, since the matches considered were the best of three sets, the match. In sport, winning contributes to further wins.

Does a testosterone feedback loop power the winner effect in humans? The pre-competition surge in testosterone has been documented in a number of sports such as tennis, wrestling and ice hockey, as well as in less physical competitions, such as chess, even medical exams. Winning athletes experience a post-game spike in testosterone, suggesting that a positive-feedback loop is indeed the physiological substrate to winning and losing streaks. Incidentally, these testosterone-driven sporting victories appear to be more common when an athlete is on home turf, the so-called home advantage. Athletes on a winning streak may thus have a very different body chemistry than those on a losing streak. In all these experiments, with both animals and humans, the winners experienced a self-reinforcing upward spiral of testosterone.

It is not surprising, therefore, that sports scientists spend a great deal of time figuring out how to raise testosterone levels in their athletes – legally, of course. We are in the curious situation of harbouring within our bodies chemicals that can unlock our full potential – kindling our competitive spirit, focusing our attention, granting access to our metabolic resources, raising us to a state of flow – yet we cannot readily access them. How frustrating! We hold the keys to victory within us, but usually we cannot find them. We would love to self-administer these drugs, but it cannot be done by a mere act of will; you cannot simply say, 'I want lots of testosterone now!' or 'I want to play my best, now!' It just does not work. Instead we must engage in all sorts of occult rituals and physical exercises before our bodies will even consider our request for more power. The situation is comparable to having money in the bank but being unable to touch it, unless, say, you perform some ceremonial dance, at which point the gates protecting your fortune swing open.

What rites need to be performed before our bodies will grant us access to our own physiological wealth? Sports scientists probably know more than anyone about what needs doing. Their techniques include altering the balance between aerobic to anaerobic exercises, between fun and gruelling sessions, their timing and length, weights lifted – in general the greater the weight, the shorter the sessions, the larger the anabolic result – diet, amount of sleep, and so on, until their athletes achieve just the right levels of testosterone. A previous victory, as we have seen, is also a potent way of unlocking our testosterone. In fact, it has been found among ice hockey players that replaying a video of a previous win can increase their testosterone levels and thus their chances of winning the upcoming game.

Another influence on testosterone levels is a rivalry: Ayrton Senna and Alain Prost in Formula One racing; Muhammad Ali and Joe Frazier in boxing; John McEnroe and Björn Borg in tennis. Often felt as a thorn in an athlete's side, the rival may bring out the best in him. After Borg retired from tennis, McEnroe confessed that there was 'this void, and I always felt it was up to me in a sense to manufacture my own intensity thereafter'. That is something many athletes do before a game: manufacture intensity, get the juices flowing. Often, without understanding the physiology involved, they try to artificially provoke the effects of challenge, even victory, before a game starts. American football players, for example, pound locker doors, and boxers strut when entering the ring and stare down their opponents. Even people in business, before and during negotiations, can prepare themselves by striking what are called 'power poses' – feet widely placed, chest out, arms crossed; or feet up on desks, hands behind head: basically poses that take up more room – and these too can raise testosterone levels. Soldiers often perform similar rituals before battle, to get the hormones flowing and to invoke Ares, god of war. In this process, music can play an important role. Napoleon complained that the barbaric music of the Cossacks drove them to such a fury that they were able to wipe out the cream of his army; and General Nikolai Linevich later said that for the Russian army, music was a 'divine dynamite'.

Away from the sports and battlefield you can feel something like a martial spirit creep through a rural population as hunting season approaches. In the autumn, throughout small towns in Canada and the northern United States, hardware shops display rifles, decoys, blinds and camouflaged clothing; grocery stores sell baskets of old apples and carrots, bait for the unsuspecting deer now grazing openly in farmers' fields; and locals begin wearing their hunting fatigues weeks before the season actually opens. In these towns you can feel a muted excitement, not unlike the animation among children as Halloween approaches, the thrill of the fright. But in these small towns verging the boreal forest, as the harvest moon wanes and a hunter's moon rises over stubbled fields, the spirit feels more like a bloodlust, and is unnerving. Come hunting season, most city folk with a cabin in the woods high-tail it out of there before the bullets start flying.

How long can elevated testosterone levels persist? Among animals the winner effect varies dramatically between species, in some lasting only a few minutes, in others a matter of months. And testosterone levels vary dramatically through the year, especially across the breeding season. Few studies exist of long-term hormone changes in humans, but the ones that do exist suggest that the rise and fall in baseline testosterone levels can be long-lasting. For example, men can experience lowered testosterone for up to six months after their partners give birth, while those re-entering the singles market after a divorce can have elevated testosterone for several years. Men living in urban areas, according to a study of the Aymara people in Bolivia, are reported to have higher average testosterone levels than people living in rural areas. In one international study, the residents of Boston were found to have significantly higher average testosterone levels than the Lese of the Congo, the Tamang of Nepal and the Ache of Paraguay. The data in these studies were based on small samples, so their interesting conclusions require further study, but they suggest that more competitive environments, such as a free market, may call forth higher levels of testosterone. In short, levels of testosterone can rise and fall for extended periods of time, even years.

There is some evidence that short-term fluctuations in hormone levels can be communicated to other people. For example, the rising and falling levels of an athlete's testosterone can be mimicked in team-mates – one Wayne Gretzky or Michael Jordan can inspire even a bedraggled assortment of players to great heights of performance. Fans are also susceptible: one group of scientists took testosterone samples from fans before and after the 1994 football World Cup final between Brazil and Italy. Both teams' supporters went into the game with elevated testosterone, but when Brazil had won their fans' testosterone levels rose, while the Italians' crashed. It appears that the runaway testosterone cycles of athletes – and the same could be true of political and military leaders – can be experienced vicariously by observers. This mechanism raises the prospect of large groups of people experiencing an upward spiral of confidence.

The literature on the winner effect, in both animals and athletes, certainly provides grounds for suspecting that a testosterone feedback loop may operate in the financial markets. Does testosterone rise with a win in the markets, and does this lead in turn to increased risk-taking? That is the question I hoped to answer. To do so I set up an experiment on the trading floor of a mid-sized firm in the City of London. The floor employed 250 traders, all but three of whom were men. They were all engaged in high-frequency trading, as described in Chapter 3, meaning they bought and sold securities, sometimes in sizes ranging up to $1 or $2 billion, but held their bets only for a matter of hours or minutes, sometimes mere seconds. They therefore occupied the same market niche as the black boxes.

These traders were therefore up against some of the world's most sophisticated and well-capitalised competitors. They lacked the large capital base and informational advantages of the flow traders at the big banks, and the deep pools of capital and inhuman processing speeds of the black boxes. Yet they were astonishingly successful: David against Goliath, John Connor against the Terminator. In fact they were some of the best traders I have ever seen: highly disciplined, consistent, and profitable.

I sampled testosterone from these traders and recorded P&L over a two-week period. What we found was that their testosterone levels were significantly higher on days when they made an above-average profit. More intriguing, though, was what we found when we looked at testosterone levels in the morning, because these predicted how much money the traders would make in the afternoon. When the traders' morning testosterone levels were high, they went on to make a lot more money in the afternoon than they did on days when their morning testosterone levels were low (see fig. 9). Moreover, the difference in P&L between high- and low-testosterone days was large, amounting in statistical terms to one full standard deviation, a difference that if annualised could amount for some of the traders to over $1 million in pay.

This was a troubling finding. Efficient market theorists tell us that the market is random, and therefore no trait we possess can affect our trading and investment returns. It should not matter how intelligent you are, how well you did at school, how thoroughly you were trained – all these have as little effect on your returns as they do on your ability to roll a dice. If that is the case, how on earth can a molecule affect the amount of money you make?

My colleagues and I found further evidence that this molecule influences a trader's profitability, and we did so more or less by

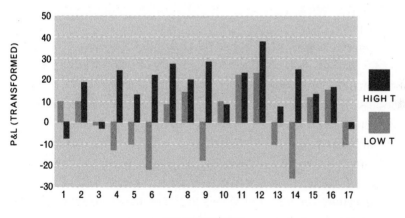

TRADER NUMBER

Fig. 9. Morning testosterone levels predict a trader's afternoon profits. Each of seventeen traders is listed along the bottom axis. Lighter bars indicate the trader's afternoon P&L when his morning testosterone was low relative to his median level during the study; darker bars when it was high. (P&L numbers have been transformed. This result can be reported more accurately as panel data. See endnote.)

chance. When I was on the trading floor conducting the first study, I brought a stack of science papers with me to read during downtime. One of these recounted an experiment in which the author, John Manning, had taken handprints from a group of football players, and found that their ability and success could be predicted from the lengths of their index and ring fingers, and specifically from the ratio of the two. This ratio, known as 2D:4D, meaning second digit length divided by fourth, could predict sporting ability because, Manning claimed, it gauged the amount of testosterone the athletes had been exposed to in the womb, a longer ring finger relative to index indicating higher androgen exposure. The idea at first seemed crazy to me, but also fun, so I began collecting handprints from traders. Later, when the back office of the trading firm sent me the P&Ls of these traders, I found to my amazement that their 2D:4D ratios predicted their profitability over the previous two years. What was even more astonishing was that their 2D:4D ratios predicted how long these traders had survived in the business. The results suggested that a hormone these individuals had been exposed to before they were even born was predicting their lifetime performance in high-frequency trading.

What is going on here? When looking into the science behind the 2D:4D marker, we learned the following. Recall the surge in prenatal testosterone, occurring between the eighth and nineteenth week of gestation? This hormone has such powerful effects on masculinising the foetus that it leaves traces all over his body, which later in life can be read off as a measure of prenatal androgen exposure, much like the high-water marks on a breakwater. 2D:4D is just one of these traces. There are others, and they are equally bizarre, like oto-acoustic emissions, inaudible sonar-like clicking in the inner ear whose frequency correlates with prenatal testosterone levels; as does ridge-count asymmetry in fingerprints; or ano–genital distance, which is exactly what it sounds like. In many hospitals ano–genital distance is now routinely measured in newborns as a way of ascertaining if they have been exposed to an abnormal prenatal steroid environment, something that can be caused by environmental hormone disruptors, in other words chemicals we put in the environment that act like oestrogens and can cause developmental problems in males, such as undescended testicles and, later in life, prostate cancer. But alas, ano–genital distance is not a marker that is readily collected from traders, although it was suggested that if we were to leave a photocopier in the middle of the office Christmas party we might end up with some samples.

Barring that option, the finger-length ratio 2D:4D has proved the most convenient measure of prenatal androgen exposure for behavioural studies. Some studies have argued that it is a reliable measure of foetal testosterone production because a class of genes, called *hox-a* and *hox-d*, code for – to use the delightful title of a science article – fingers, toes and penises.

A vital question remains: how is testosterone having its effects on P&L? It may be that testosterone is affecting the traders' performance by increasing their appetite for risk or their confidence. Equally, it may stabilise their visual attention, reduce distractions from irrelevant information, maintain search persistence, or augment visuomotor skills such as scanning and speed of reactions, thereby permitting traders to spot price anomalies faster than their competitors. We will not know which of these aspects of their skill are affected until we have completed more controlled studies in the laboratory.

But we did make some headway in answering this question in the previously discussed study that looked at traders' Sharpe Ratios, in other words how consistently they made money, their P&L corrected for the amount of risk they took in making it. In this study we used the traders' Sharpe Ratios as a measure of skill, and asked quite simply, does testosterone improve traders' skill, or does it increase the amount of risk they take? What we found was that testosterone did not improve their Sharpe Ratios, but it did increase the risk they took. We retain the belief that testosterone also has effects on the traders' visuomotor scanning and speed of reactions, although we will not be able to test this in the field. What our trading-floor studies did highlight, however, was that testosterone, a signal from the body, was having a large effect on the risk-taking of traders, rising with above-average profits and increasing risk-taking. These experiments thus provided good preliminary evidence that the winner effect does in fact occur in the financial markets.

EXUBERANCE

In the next few days Scott doubles his core position to 4,000 S&P contracts. At this point warning bells go off in the risk-management department of the bank, and Scott is asked to justify his decision. He takes the risk managers through his logic, as he has already done with Stefan and Ash. The risk managers understand his reasoning, may even agree with it, but they still have grave concerns. What happens in a crisis? What happens when the market begins to worry about credit risk? Right now everyone seems blissfully unaware of that danger.

Risk managers are usually pretty sharp-eyed about these things, often being former traders themselves – send a thief to catch a thief – and possess an enviable competence in statistics. But, alas, they lack clout. At the end of the discussion it is Ash and the desk managers who have the final say. And here, unfortunately, we find management practices and compensation schemes magnifying the biological forces that push traders to take more risk. Scott has made money for the past few years, and this year he is on a roll. Why rein him back, managers ask themselves. More decisively, where is our interest in doing so? Our year-end bonus is calculated as a percentage of our P&L, both for individual traders and for managers, so we want to make as much money as possible in this calendar year. Who cares if our positions or strategies blow up next year? We don't give back previous bonuses. So it is in Scott's, and his managers', interest to take large risks. In fact, there is a subtle pressure on trading floors to constantly engage with the market, to take risks. I once went through a period of inactivity, pacing the floor like Hamlet, racked by indecision – to trade or not to trade – and my boss, failing to appreciate the depths of my existential angst, told me quite simply, 'Coates, shit or get off the pot.'

Scott receives permission to increase his position, and one day when the market dips 1.5 per cent on bad economic news, he buys another 1,000 contracts. With a core position of 5,000 contracts he makes or loses $3.25 million if the market moves up or down 1 per cent – or $32.5 million if it drops 10 per cent. The sheer magnitude of the risk tinges his waking hours with fear, but below that, more powerful, simmers a confidence, an unshakeable belief in his ability to dominate the world. Exercising his skill in this way, Scott against the market, makes him feel keenly alive, and his life seems to accelerate. Like an adolescent discovering his strength, Scott flexes his new powers, from his brain right down to the ends of his fingers. His mind, quick and lithe, moves from thought to thought effortlessly, although to outsiders it appears he cannot keep his thoughts on track; he needs less sleep, and the potent cocktail of dopamine and testosterone being stirred in his brain fills him with a life-enhancing euphoria.

Others on the desk catch the mood, place copycat trades, and together they revel in an ecstasy of risk. Ash escorts a new saleswoman down the aisle to her desk, and behind her, just shy of her field of vision, a human wave follows as arb traders, peering over screens, check her out.

Normally these moments of profit and cockiness build, crest and wane. But not in a bull market. There is no downtime. Traders get on a roll and stay there, and under such circumstances their physiology does not have a chance to return to normal. It is at this point that they enter what may be called the end-game of the winner effect.

IRRATIONAL EXUBERANCE

What an extraordinary mechanism of empowerment the winner effect is. By its means a single person could conquer the world, or so it must feel. How far can these feedback loops run? Not surprisingly, they cannot go on forever. Biologists have found that the effects of testosterone on risk-taking among animals display the same ∩-shaped dose–response curve that we have encountered before, in Berlyne's hill for example. At low levels of testosterone an animal will lack motivation, arousal, energy, speed and so on, but as testosterone levels rise so too does the animal's performance in competitions and fights. When testosterone reaches its high point on the curve the animal enjoys optimal performance. It is in the zone. However, should testosterone continue to rise, the animal begins to slide down the other side of the hill, and its risk-taking becomes increasingly foolish. Male animals experiencing a sustained rise in testosterone tend to start more fights, patrol larger areas, venture into the open more, and neglect parenting duties, all of which lead to increased predation and reduced survival. At some point, as testosterone builds up in these animals, confident risk-taking morphs into overconfidence and rash behaviour.

Equally powerful psychological effects have been documented among athletes and recreational users who are 'ripped' on anabolic steroids. Pope and Katz, two psychiatrists from Harvard, found that many of these people succumb to mania, a psychiatric disorder in which the patient becomes euphoric and delusional, and experiences racing thoughts and a diminished need for sleep. In one case, a university athlete on steroids, after buying a sports car he could not afford, became so convinced of his invincibility that he asked a friend to film him driving the car into a tree, to prove he could not be hurt. In other cases, users of steroids have committed criminal acts, and in their defence have blamed their behaviour on testosterone, a legal strategy that has come to be known as 'the dumbbell defence'. We have to interpret these cases with caution, because the athletes were taking steroids at levels far higher than can occur naturally in our bodies. Yet their behaviour is not unlike what I observed in many traders during the dot.com bubble.

There is a further cost of high testosterone. Elevated testosterone, and the larger or more ornamented body it promotes, is energetically expensive, and can eventually wear down an animal's body. Castrate a male animal and it can live up to 30 per cent longer. High-testosterone males thus end up paying a high price for their show of strength and their triumphs, in the form of a higher rate of mortality. It has been said that there is a certain tragic glory to these highly charged males – 'The candle that burns twice as bright burns half as long.' Achilles and Macbeth, it might be said, did not suffer at the hands of the gods; they paid the price for high testosterone. Today something of this tragic spirit still hangs about the figure of an extreme male, fighting to assert himself against ultimate failure. It is almost as if they know they are doomed. Many men sense that beyond the sports field, the battlefield, and perhaps the trading floor, testosterone no longer plays a useful role in the workplace or in society. They feel it is just a matter of time before age and the gods of the service economy crush them.

In the financial world, testosterone feedback loops, once they start, can cause traders to pass through the early stages of thrill and excitement, and end up convinced of their own infallibility. As these cycles rise to their euphoric high point, one finds traders, most of whom are young males, with impaired judgement, doing dangerously silly things. Following the pattern of the winner effect, traders experience a rise in testosterone when their trades make money, which increases their confidence and appetite for risk, so that in the next round of trading they put on even larger trades. If they win again, as they are likely to during a rising market, their profits will increase their testosterone once more, until at some point confidence becomes over-confidence, trading positions grow to a dangerous size, and the risk–reward profiles of the trades start to stack the odds against them. But no matter; in their overconfident state, traders are convinced they will win anyway. As is management. When a trader makes more and more money, managers expand his risk limits apace. As a result traders are walking time-bombs, and banks invariably light the fuse, dangling before them huge risk limits and bonus payments that have exceeded $100 million. No wonder the traders responsible for bank-crippling losses frequently turn out to be the stars of yesterday. Banking is an odd world. I know of no comparable behaviour among surgeons or air-traffic controllers.

Rising testosterone, it should be pointed out, does not start a bull market; usually a technological breakthrough or the opening up of new markets does that. But testosterone may be the catalyst that turns a rally into a bubble. It may by the same token be the chemical encouraging other expressions of overconfidence one sees towards the peak of a bubble, like oversized and ill-considered corporate take-overs, or the construction of record-breaking skyscrapers, such as the Empire State Building, commissioned at the end of the Roaring Twenties, or the Burj Dubai (renamed Burj Khalifa after it became insolvent), built during the recent housing bubble. Testosterone may be the molecule of irrational exuberance.

HIGH SEASON

In the weeks after Scott adds to his position, the stock market proves particularly volatile. His P&L now swings on a daily basis more than it ever has before. Up $4 million one day, down $3 million the next, up $7 million, down $5 million. Scott finds this volatility bracing, as it enables him to display to all that he can absorb these blasts and remain on his feet. On one especially bad day, when facing a loss of almost $8 million, he shows lesser mortals the full extent of his courage by adding another 1,000 contracts to his position, taking it up to the full limit agreed by Ash and the risk managers, 6,000 contracts – and this in addition to the large spread trade he has maintained for months now between stocks and bonds. He has jumped on this weakness in the market because tomorrow sees the release of statistics on the state of the housing industry. The market is worried about these numbers, because house prices have been dropping for months now, mortgage foreclosures have risen, and buyers of new homes have for the moment disappeared. Yet all this is music to Scott's ears, as it makes it easier for him to add to his trade.

This is a crazy bet, executed in insane size, with a terrible risk–reward trade-off. Stocks are already too expensive by historic standards, but to add to this position when the housing market weakens is sheer folly. Cooler minds on the floor, like Martin and Gwen, hearing of Scott's position, exchange knowing glances. A bull market, like a river in spate, carries almost all before it; and the few people on the floor who do not buy into the frenzy start to feel like outsiders. The atmosphere can be compared to that of an exciting party, bubbling with possibility and animated conversation, but when you listen to the talk you find you cannot keep up, or get a word in, or even see where the interest lies in the things being said; and then it finally dawns on you: you and a few others at the party are the only ones not coked to the gills. That is what it feels like in a bank during a bubble. The few people left unaffected by the narcotics whisper in coffee rooms and after hours about the insanity that could blow up their bank; but the rest of the crowd are beyond reach. No amount of statistics, no history of price–earning ratios, no reasonable chat can bring their faraway look down to earth. To them, the money being made holds out the promise of too many things only dreamed of: a penthouse on the Upper East Side, a private jet, even political clout – My God, I could be a player! – all this is right there for the taking. Such people are caught firmly in the delusional phase of the winner effect.

The next morning, however, brings a nasty wake-up call. The economic statistics show a worsening trend in the housing market, with sales down 3 per cent and prices nationwide sliding 2.5 per cent. Stocks promptly drop 1.5 per cent. But the fear does not last. The market smiles on Scott and turns an upward thumb. A weak housing market means without a shadow of a doubt that the Fed has finished raising interest rates for this business cycle. In fact it may now be forced to lower them, and the very possibility acts as a flame to a market already soaked in fuel.

Investors during bubbles seem to come equipped with special eyewear that permits them to view all economic news as bullish. Weak economic growth spells lower interest rates, so stocks and risky assets rally; strong economic growth means healthy balance sheets for both corporations and households, so stocks and risky assets rally. The dollar strengthens, and this means foreigners love US assets, so they rally; the dollar collapses, and this aids exporters and hence the economy, so assets rally. With this kind of spin no news dampens animal spirits for long, so by noon the stock market, celebrating the impending lower rates, has started to rally, and rally strongly. By the end of the day the S&P stands 3 per cent higher. Scott is mesmerised by his risk-management system, hooked electronically to the screens, for it calculates in real time the P&L on both his S&P contracts and his stock-to-bond spread trade; and towards the end of today not even he can remain calm as the number climbs to almost $15 million, bringing his year to date close to $32 million, more than he has made in his best full year.

The news transports Scott into another reality. Every one of his predictions has come true, every trade he puts on makes money. It takes a sober person not to be affected by relentless triumph of this magnitude, and Scott was never very sober to begin with. He, along with a few others, has today ascended into the realm of masters of the universe. There is nothing he cannot do. Rumours spread of his big win, and traders and salespeople timidly peek at the new hero. Scott breathes in the floor, hears its tumult and varying fortunes – the sound of the market itself, and all earthly glory. After acknowledging the adulation of his peers and the hints of a managing directorship that he takes as his due, he struts the main aisle of the trading floor, known appropriately in many banks as Peacock Walk, and heads out into a New York evening sparkling with opportunity.

In ancient Rome, when a general had achieved a great victory he was awarded a Triumph, a ceremonial parade through the centre of the city. But the ancients were clever; to prevent the general's hubris from ruining him, they placed in his chariot a slave whose job it was to whisper in the general's ear a reminder that he was not a god. 'Remember this,' the slave warned, 'you are mortal.' To bring home the point he would hold in the general's line of vision a human skull, a *memento mori*, a vivid sign of his inevitable fate. But alas no such *memento mori* exists within the banks, so very little tethers a winning trader to earthly standards of prudence.

In the coming months, Scott's P&L breaks new ground, amounting to almost $45 million, a figure which could well mount to $60 million by year end, putting him in line for an $8 million bonus. Despite his winning streak, Scott decides to close out his 6,000 stock index contracts. This decision does not stem from an uncharacteristic moment of prudence. Scott still believes in the trade, still thinks the bull market has miles to run, but he unwinds some of his risk for one

182

very important reason. August is approaching, and that means it is time to move the carousing from Wall Street to the Hamptons, where Scott can play and party and brag with the other trading heroes. Despite a steady stream of statistics portraying a worsening economy, which Scott and others have managed to ignore, he leaves his spread position – stocks to bonds is a money machine, so let it keep printing – to be tended by assistants.

On the drive out on the expressway, as the sprawl of Queens and lower Long Island gives way to the pines and orange sand and the intensifying drone of cicadas, Scott sheds his workday concerns. High summer. The still point of the year. Scott basks in the knowledge that this is one of the last summers he needs a rental, the last time he shares. In the coming years he will own his own house on the beach, one of those half-timbered beauts from the 1920s about which there hangs an aura of almost otherworldly exclusivity.

SEVEN

Stress Response on Wall Street

Occasionally, it seems, when the world drifts unknowingly to the edge of an abyss, nature conspires to prolong a particularly glorious summer, as if to forestall the impending disaster, or to heighten the irony future historians will read into the prelude. Take, for example, the idyllic summer of 1914, coming at the end of the elegant and oblivious years, known nostalgically as the Edwardian Summer, leading up to the First World War. Or the New York autumn in 1929, when a heat wave lingered after vacationers had returned from the beach.

And so it was this September when Martin and Gwen and Scott and Logan, tanned and toned and ready for the final run to bonus time, straggled back to work, despite an Indian summer that would not relinquish its calm seas and golden days.

RISK ON, RISK OFF

For Logan, though, the summer break had not been quite as carefree as he had hoped. He was repeatedly called off the water by his wife waving his cell phone in the air, and would then have to spend an hour or two talking with the trading desk. For the market he trades, the market for mortgage-backed bonds, spent the month of August on a Coney Island-like rollercoaster ride as worries about the credit-worthiness of home-owners and mortgage lenders rose and fell with ever increasing intensity.

Mortgage-backed bonds consist of a large number of individual home mortgages that have been bundled together and issued as a single bond – securitised, as bankers call it – and sold to investors. They differ from Treasury bonds in a number of ways: if you buy a ten-year Treasury yielding 5 per cent, investing, say, $10,000 of your savings, you will get yearly interest payments of $500, and at the end of ten years you will get your $10,000 back. All of it. You have the US government's promise on that (although recently a few people have questioned that promise). Mortgage bonds, on the other hand, are paid back by home-owners: you get yearly interest payments as home-owners make their mortgage payments, and in theory you get your money back at the end of ten years, if the mortgages underlying the bond mature at that date (most American mortgages mature in 30 years). You may get all your money back in ten years; or you may get very little of it back if the home-owners cannot pay back the money they have borrowed. For this reason, mortgage-backed bonds expose you to a greater risk of losing money than Treasuries, and consequently need to offer higher yields to entice investors to buy them. For much of this year they have had to offer about 6.10 per cent interest, a full 1.10 per cent higher than the 5 per cent on ten-year Treasuries. At this level they have tempted investors; and whenever mortgage bonds have yielded more than 1.10 per cent over Treasuries, investors have quickly stepped in to buy them.

During the spring and summer, however, the pattern changed: the mortgage market weakened, offering higher yields, yet the usual buyers failed to show up. Rumours circulated about home-owners getting into financial trouble and being unable to make their monthly payments; foreclosures on homes rose rapidly, and a couple of large mortgage lenders declared bankruptcy. Investors rightly began to fear they might not get back the money they had lent to home-owners. As a result yields on mortgage bonds have risen, and now offer a full 1.60 per cent more than Treasuries, making them more attractive than they have been in years.

August had provided some scary moments, but by mid-September the panic in the mortgage market has largely subsided. Traders ponder the higher yields on mortgage bonds, and find them attractive. Scott in particular thinks they are a steal at this price. When he and Logan are not tossing a tennis ball back and forth they are bouncing trade ideas off each other, and during one of these brainstorming sessions they decide to reload on credit risk. Logan opts for a spread trade by buying mortgage-backed bonds and shorting Treasuries as a hedge. He expects mortgages to increase in price relative to Treasuries, but – and this is the real attraction of the trade – while he waits for this to happen he is receiving 6.6 per cent interest on his mortgages but losing only 5 per cent on the Treasuries he is short. In other words, he is collecting interest almost for free. It is a very attractive position, so when investors and traders lose their fear of home-owners defaulting they rush into these spread trades, driving the price of mortgages up relative to Treasuries.

What happens on the mortgage desk happens on all desks across the trading floor that trade credit-sensitive securities. There too, traders buy corporate bonds, junk bonds, bonds of emerging market governments, and sell Treasuries against them. Scott for his part has re-established his long position in stocks. The trouble is, when crises erupt, all these trades have a nasty habit of blowing up together. Their performance waxes and wanes with the appetite for credit and risk in general. Lately, journalists have termed this seesaw in the financial markets 'risk on, risk off'.

By the middle of September, it is risk on. After several years of a roaring bull market, enthusiasm is not easily dimmed by a few bad stats out of the housing market. In fact, for the past ten – no, twenty – years every scare in the market, every sell-off, has proved to be a buying opportunity, and that is how traders perceive this one. Old buying habits reassert themselves, and over the next few weeks the S&P 500 surges to an all-time high. Credit spreads once again contract, mortgage bonds rallying almost $3 relative to Treasuries.

Emboldened by the money they have made on stocks and mortgages, Scott and Logan now think about increasing their exposure to the credit markets, and look in particular at one segment of the mortgage market that has not recovered as much as the others. This is the market for sub-prime mortgage bonds. Mortgages are considered sub-prime if they are taken out by people who may have a hard time paying back the principal or indeed even making their monthly payments. For that reason these bonds have to offer considerably higher yields to investors. Instead of 6–7 per cent interest, they offer, depending on the risks of default on the underlying mortgages, a tantalising 10–15 per cent.

Scott cannot resist yields this high. He decides to sell his long position in stocks and buy some of these distressed bonds instead, thinking they represent better value. However, because they often prove illiquid to trade, meaning he cannot easily buy and sell them, he decides on buying an index called the ABX, which tracks the average price of a basket of sub-prime mortgages, just as the S&P 500 index tracks the average price of 500 stocks. The index he decides on was originally issued at a price of 100, but has now dropped to a bargain-basement price of 41, a loss of 59 per cent. Scott buys what he considers a modest amount of this index, $300 million. If the bonds drop to 37, which he thinks is a worst-case scenario, he could lose about $12 million, but he thinks it more likely the bonds will rally to 55–60 over the coming weeks.

Logan hates to be left behind on opportunities like this, so he too buys sub-prime mortgages, but only $100 million. He is already long mortgages through his spread trade; and in his capacity as a flow trader of mortgages he finds that he is continually buying them from clients, frequently more than he wants. Logan believes in the trade, but he does not take a bigger position, even though he has frequently done so, mainly because it is October and the fiscal year is drawing to a close. His P&L year to date shows him up a comfortable $29 million, putting him in line for a nice bonus – maybe $4 million. So why risk his P&L this late in the year? January is the time to start taking big risks again, because if they go wrong you have the rest of the year to make back the money. Late autumn, most traders agree, is the time to cruise to the finish line.

The next couple of weeks, moreover, may prove to be a particularly dangerous time to put on a large mortgage trade, and Logan knows it. There are scheduled for release a number of economic reports that the market has been anxiously awaiting: those showing US new home sales, the Case-Shiller house-price index, US GDP, and then the latest Fed meeting. All this information will shed light on a housing market many fear is hollowing out the American economy.

Unfortunately, in the coming week the traders' fears turn out to be fully justified. The news is indeed bad. In fact it is worse than bad – it is horrendous. Sales of existing homes have dropped almost 8 per cent in one month, and the Case-Shiller index shows its largest drop ever, with house prices nationwide falling 8.5 per cent. The news shocks a market already reeling from record defaults on sub-prime mortgages. Scott and Logan start to lose money on their mortgage position almost from the day they put it on, and soon the ABX index is trading at 37, already reaching Scott's worst-case scenario of a $12 million loss.

Logan is furious at himself for losing so much money at year end, and terrified by the way mortgages are trading. Through the sales force he sees nothing but selling, so he closes out his sub-prime position and concentrates on trying to hedge all the mortgages clients keep selling him.

Scott, however, does not despair, because other markets appear much more sanguine about the economic statistics. Stocks and corporate bonds take the news in their stride, trading off a bit, but basically holding firm. Nervous traders, Scott included, interpret this solidity to mean that the worst of the news may now be over; the economy should start to improve. Such an interpretation is strengthened the next day, October 31, when the GDP report shows that the US economy continues to grow at a strong pace, despite the slowdown in housing. To top it all, the Fed lowers interest rates by a quarter point, and states in its press release that 'economic growth was solid in the third quarter, and strains in financial markets have eased'. All this spells one thing to traders raised on the milk of optimism – the bull market is back. Stocks rally back to their highs, and Treasuries sell off as investors feel emboldened to venture from their safe haven. Mortgages do not participate in the rally, which is disconcerting, but given the strong growth in the economy and the Fed's confidence, traders feel it is just a matter of time before this market too recovers its pluck. Relief is palpable across the trading floor.

Towards the end of the day, the excitement that has been building in the market transfers to Halloween celebrations and the evening ahead. Pumpkins are put out on one or two desks, a tombstone on another, and people look forward to trick-or-treating with their kids or heading out to costume parties, Manhattan at this time of year looking much like a film set for *Night of the Living Dead*. But just after the market closes, before the fun starts, something quite unexpected happens. Wall Street gets an early scare as an unscheduled announcement trickles across the news wire. An analyst at a Canadian investment bank has downgraded its assessment of Citibank's future earnings, and in a strongly-worded statement cites Citi's large inventory of bad mortgage loans. One analyst, one bank – Canadian at that. So what? Yet the news spooks the market in the after hours, and nags at it all through the night.

What is it about October, anyway? Why is it always the scariest month for stocks? Almost every crash in the history books, at least those occurring in the US and UK, has taken place in the autumn, and most of those in October. The Panic of 1907, the Crash of 1929, Black Monday of 1987, the Crash of 1997 (related to the Asian Financial Crisis), the crashes of both 2007 and 2008 (related to the Credit Crisis) – all took place in October. It was thought in the nineteenth century and the early twentieth that crashes occurred in the autumn because farmers, needing cash for the harvest, withdrew their money from banks, causing bank runs and stock crashes. Perhaps that pattern lingers in our collective unconscious. But I will throw out another possibility. In many animals testosterone levels fluctuate over the course of the year, and in humans these levels rise until the autumn, and then fall until the spring. This autumnal drop in testosterone can lead animals into a condition called 'irritable male syndrome', in which they become moody, withdrawn and depressed. So maybe, just maybe, in the autumn traders' animal spirits give up the ghost and risk-taking dims, taking stock markets down. While thinking along these lines, I might point out another oddity in the stock markets, and that is their observed tendency to outperform on sunny days and underperform between the autumn equinox and the winter solstice, an effect some have attributed to Seasonal Affective Disorder. Perhaps this too can be traced back to testosterone levels, for these increase with sunshine, and decline during autumn and winter months. Food for thought.

Next morning, as traders straggle back to work, many of them hungover, the market feels different. It is as if the accumulation of bad news, each piece on its own dismissible, has reached a tipping point. Stocks start the day in an ugly mood, Treasuries are in demand once again, and mortgages look wobbly. Shen goes over the squawk to give some colour on the news, and states the bank's view that the housing market is in bad shape, but exports and manufacturing are holding up, so he does not expect too much damage to GDP. Martin follows Shen, but says he expects Treasuries to keep rallying, as the credit problems are only just starting. A nervous tension ruffles the floor. Traders are on edge, vigilant, sensitised to bad news.

And then it comes. Shortly after Martin's commentary, two more announcements scroll across the screens. Traders and salespeople freeze as they read the news. Analysts at Morgan Stanley and Credit Suisse have also downgraded Citibank, and confirm the extent of the damage it has suffered. It is at this moment, all along Wall Street, that the penny drops. The US housing market is falling off a cliff, and is taking the banking system with it. No one on the floor, not even the most pessimistic of bears, anticipated anything like this. A momentary beat ... then a roaring, desperate bedlam. Shen tries to comment over the squawk box, but is drowned out by salespeople from around the world shrieking for bids on mortgage-backed bonds, on corporate bonds, on anything with credit risk; others plead with Martin and Gwen for offers on Treasuries, in large size. Martin, his vagal brake still firmly engaged, pauses before quoting any prices, and watches as the mortgage market drops like a knife while the ten-year Treasury rallies half a point. Once he has a fix on prices Martin, like a skilled air-traffic controller, rapidly deals with a long line of waiting clients. As volatility rises, Martin and Gwen take centre stage as two of the only money-making traders on the floor.

Scott, on the other hand, stares at his screen, stunned. He is long sub-prime mortgage bonds, and they have just dropped $2, without any trading on the way down. He has just lost $6 million, and that on top of the $12 million he has bled over the past week. Deep in his brain, ancient circuits register the anomaly and the fact that it is a nasty one. From this point on, Scott's body and brain begin to undergo far-reaching changes as a vast network of electrical and chemical circuitry switches on. For his amygdala, the emotional centre of the brain, has tagged this event as particularly dangerous, and has triggered the initial phase of what is called the 'stress response'.

FIGHT-OR-FLIGHT ON THE TRADING FLOOR

The stress response is a rapid switch in body and brain away from everyday functions to a state of emergency. It evolved to deal with imminent physical threats, such as an accidental encounter with a mountain lion while foraging in the woods. In preparation for an exceptional muscular effort, be it fighting for our life or sprinting to safety, our body marshals all the glucose and oxygen it can, while shutting down long-term and metabolically expensive functions of the body. The stress response is an overwhelming experience, and over the long sweep of evolutionary time it has proved essential in keeping us alive. However while the stress response is useful when we are faced by a mountain lion, it can prove largely counterproductive when seated in the workplace. Indeed, workplace stress provides a vivid illustration of how our body can have a plan of its own for handling a crisis, one over which our conscious minds have little control.

The stress response unfolds in several stages: two fast ones, employing electrical impulses, and two slow ones, employing hormones. First, the amygdala must register the danger and pass on a warning via electrical signals to other parts of the brain, a rapid process taking place in a matter of milliseconds. Second, electrical signals sent from the amygdala via the brain stem to visceral organs such as the heart and lungs increase our heart rate, blood pressure and breathing. These signals begin to have their effects in less than a second, although their full effects can take a little longer. The initial electrical responses in body and brain are thus lightning fast, and when successful carry us clear of danger. But they are metabolically draining, and burn out quickly without more fuel. This fuel is provided by slower hormonal responses, such as adrenalin, which works its effects over the course of seconds and minutes. These early stages of the stress response constitute the 'fight-or-flight' response. This response, as we have seen, is initiated by any situation requiring a quick mobilisation of energy and attention. The hungry wolf chasing an elk, and the terrified elk running for its life, experience much the same fight-or-flight response. For that matter, so do Martin and Scott, even though one is in control and the other is not. In that respect Martin and Scott are not unlike wolf and elk, predator and prey.

Martin's and Scott's physiologies differ, however, in the final stage of the stress response. If a crisis lasts longer than fight-or-flight, then the shell of the adrenal glands, called the adrenal cortex (cortex means outer layer), secretes ever increasing amounts of cortisol. This hormone, the big gun of the stress response, brought into action to support us in a more sustained effort, takes effect over the course of minutes to hours, even days. Cortisol has powerful effects on our brain, and our health, and while Martin experiences moderate increases in this hormone and benefits from its invigorating effects, Scott comes to suffer ever higher levels, and these impair his judgement.

Let us watch each of these stages of the stress response unfold as Scott reacts to his money-losing position. First, he needs to become aware of the danger facing him by processing the information pouring in through his eyes and ears. One of the first brain regions to help him do so is called the thalamus (see fig. 10), found roughly at the cross-section of lines projected inwards from his eyes and ears. The role of the thalamus is to format sights and sounds as they enter the brain, so that they can be interpreted, just as data must be formatted before a computer can read it. Importantly, the formatting done by the thalamus is quick and dirty, producing what looks like a blurry, half-developed picture or what sounds like unarticulated gibberish. The thalamus then sends a rough image, in the case of a visual cue, up to the sensory cortex, where it is developed further, so that the image comes into focus and can be analysed rationally. However, at the same time as the thalamus is doing this it also passes a rough image to the amygdala, where it is assessed in a quick and tentative way for emotional significance – *Is this an image of something I like? Is this something I should be scared of? Should I be happy, sad, scared or angry?*

Why would we want to assess the emotional significance of a thalamic image we can barely make out? Because it is fast. As we have seen, our brain faces an inevitable trade-off between speed and accuracy, and in an emergency we choose the speed of pre-conscious processing. If we spot a dark moving object while on a hike through the woods, it could be a shadow formed by swaying leaves, or it could be a bear. Our rational brain will, with time, establish which of these it is, but this takes precious seconds, and if it is a bear that extra time may mean the difference between a narrow escape and none at all. Thus our brains have evolved what Joe LeDoux has called the high and low roads for information-processing: the thalamus–cortex circuit being the slow but accurate high road; the thalamus–amygdala circuit, which cannot distinguish between a shadow and a bear, the fast, low road. With the aid of the low road we react first and calm down later, feeling slightly foolish, in the case of a false alarm, that swaying leaves startled us so much.

So when the shocking analysts' reports flash up on the news screens, Scott's first amazement is processed by his amygdala, which mutely, stupidly registers: this is bad. His amygdala then passes on the bad news to the locus ceruleus and the brain stem. The brain stem, roused by the amygdala's clarion call, now accelerates a fight-or-flight response that had already been activated, albeit at low levels, before the housing number was released. Let us recap what happens and add more detail.

Electrical impulses race down Scott's vagus nerve and down nerves in his spinal cord, branch out into his body and stimulate his respiratory and cardiovascular systems. His heart rate speeds up, and with it his blood pressure, pumping the extra blood needed to fuel a fight to the finish or a dash through the forest. The surge in blood flow is selectively targeted, with arteries to the skeletal muscles dilating, forcing more blood to major muscle groups in the thighs and arms. At the same time arterioles – tiny arteries – in the skin constrict, to reduce bleeding if injured, giving Scott's skin a clammy feel and his face a pallor. Blood vessels in the stomach also constrict, since digestion is not currently needed, giving him the butterflies. Breathing accelerates as the lungs try to provide enough oxygen for the increased blood flow. The skin starts to sweat, cooling Scott's body even before the expected physical exertion begins; as do his palms and the soles of his feet, perhaps a throwback to an earlier evolutionary period when escape involved scurrying up vines or branches. Pupils dilate to take in extra light. And salivation stops, to conserve water, giving Scott a dry mouth. In cases of extreme fright the erector pili muscles at the base of body hairs can contract, causing hair to stand on end, or, where hair no longer exists, causing goosebumps.

Many of these physiological changes occur so quickly that Scott's consciousness is left behind, playing little part in his body's first reaction. After a moment or so his rational brain catches up, and unfortunately it confirms the amygdala's fast and dirty assessment – he is indeed in a pretty bad situation. At about the same time, the hormonal phase of the fight-or-flight response has started with the release of adrenalin. As adrenalin starts to course through his blood vessels it taps into the energy stores needed to support his fight-or-flight response, mostly by breaking down glycogen (the molecule used to store sugars) from the liver and turning it into glucose. The adrenalin also increases blood coagulation, so that in the event of injury his blood will clot quickly. As a further safeguard against injury, the immune system floods natural killer cells into the bloodstream in order to battle any resulting infection.

Scott needs to think clearly about his position and the market, but oddly, inappropriately, his body has atavistically prepared him to fight with or run away from a bear. The stress response is prehistorically hamfisted in this regard. It does not distinguish very clearly between physical, psychological and social threats, and it triggers much the same bodily response to each one. In this way the stress response, so

194

valuable in the woods, can prove archaic and dysfunctional when displaced onto the trading floor, or for that matter any workplace. We need to think, not run.

So far, Scott's stress response, although slightly uncomfortable, has not seriously impaired his ability to deal with his loss. Dropping $18 million at year end is definitely bad news, but Scott has lost a lot of money before and made it back. Years of trading have tempered him into a durable risk-taker, and at moments like this he proves he can resist the ancient and insistent pressures exerted by the stress response and trade effectively.

A WORSE DAY ON THE MORTGAGE DESK

Risk managers now mill about the arb desk and look over Scott's shoulder. In the background, from all across the floor come the angry sounds of thwarted plans – strangulated screams, bellowed obscenities, smashed phones. Logan has been particularly hard hit, and is in mid-tantrum. Stefan, the head of the arb desk, dealing with his own loss on derivatives, totalling almost $60 million, calls the group together for a hasty conclave. Do they close out their positions and limit their losses at year end? Or add to them, hoping to make back all they have lost, maybe more? The traders, in half-finished, telegraphic sentences, bounce opinions back and forth, agreeing the move is overdone. Hedge funds, they reason, have been shorting mortgages and will want to cover after a move of this magnitude; and besides, with the housing market collapsing the Fed will surely continue to lower interest rates, a policy that normally causes mortgages, stocks and other credit-sensitive markets to rally. Expecting a bounce any time now, the traders decide to add to their positions. The risk managers look concerned, remembering similar arguments made during the Asian Financial Crisis and the Russian Default, but agree, based on the desk's track record this year, to let the traders increase the size of their trades. Once again the risk managers are in an impossible situation – if they refuse, and the market goes up, they will get blamed for lost profits.

Scott rolls his chair back to his desk and pulls up a real-time charting tool on his computer. These charts, it is claimed, help traders find patterns in the bewildering zigzags of securities' prices. In particular, charts supposedly show what are called 'support levels' in the markets – mortgages, stocks, currencies, whatever – price levels at which investors are expected to step in and buy, driving prices back up. The charts, as many have pointed out and as Scott fully knows, are drawn and sold by people hailing from dubious intellectual backgrounds: the support levels are supposedly based on Fibonacci numbers, a perfectly respectable mathematical progression found in natural phenomena such as the spiral pattern of a seashell, but these sequences have become, disconcertingly, a staple of pop culture, turning up in novels such as *The DaVinci Code*, where they provide a frisson of hidden patterns. These charts border on number mysticism, yet if enough people believe them they become self-fulfilling. Scott, aware of how many traders follow the charts, puts in a bid to buy another $200 million of the ABX mortgage index at a price of 34.00, the next major support level, which would take his total position up to $500 million. As the mortgage market grinds its way down, bringing his daily loss up to $9 million, Scott gets filled on his buy order. Glancing at the other arb traders, he guesses they too have now added to their positions.

The stock market quickly dips another half a per cent, and mortgages drop in sympathy, surprising the guys with a grim moment, but then stabilise and creep back up to the levels at which Scott has just bought. And now Scott, together with the arb desk, and behind them the thousand-strong trading floor, and behind it the hundreds of similar trading floors around the world, waits for that reassuring elevator-tug as the markets bounce. And for a while they do, stocks, mortgages, corporates, tentatively building up confidence, gathering buyers, the incipient rally picking up steam after a rumour spreads that the Fed will make an announcement later that day, no doubt stating its resolve to support the market. Mortgages rally 2 per cent, reducing Scott's loss. Legs twitching, egging the market on, Scott feels the old magic returning, and senses relief along the arb desk. If this market keeps rallying, he could actually make money today.

But after an hour of slogging upward, cent by laborious cent, the rally looks unconvincing, tentative, and fails to build on itself. Other rumours begin clouding the picture, about mortgage originators having liquidity problems, hedge fund losses, banks taking massive write-offs on their bad loans, the British banking system collapsing, and the rally stalls. Soon the selling re-emerges – a mutual fund in the Midwest, a hedge fund in Zurich, the night desk at the Bank of Japan, the squawk box crackling with salespeople looking for bids on mortgage and corporate bonds. Confidence falters and the market fails, dropping slowly, insistently, then building up speed, slicing back through the 34 support level, through 33.75, plummeting in the next 45 minutes to 33.05. Scott was quick to sense the change, and has been trying desperately to sell out his position, but bids just disappear before he can hit them, and within half an hour his loss has grown to maybe $16 million, almost a third of his year. Even then he can't be certain of his P&L, mortgage prices jumping around so much no one can be sure where they are. Nor for that matter of his true position: he has traded so fast and frantically he cannot be certain all his trades have been entered, or entered correctly. And now, in a daze, a panic, rancid sweat seeping from his pores, Scott watches spellbound as the bottom falls out of the market and it is sucked into a death spiral, the occasional price flickering on the way down – 32.50, 32.15, 32.27, 31.90, 31.35 – a wretched din from the trading floor heard as in a dream, and by mid-afternoon, as prices settle, the news tape reports the ABX down a record 12 per cent. The back office confirms Scott's position and he is still long most of his bonds, about $415 million. Prices and positions in his risk-management system stabilise, and reluctantly, fearfully, he looks at the P&L number posted in the bottom right-hand corner of his computer screen, gasping as he sees the loss of $24 million, almost all the money he has made in the past six months. The news goes off like a depth charge in his brain, breathing accelerates, blood pressure rockets, and his bowels liquefy.

When caught in a terrifying event, our bodies assume we need a quick sprint to safety, and accordingly jettison excess weight by forcefully expelling urine from the bladder and faeces from the colon, the faeces being loose and watery. Normally, when undigested waste matter leaves the small intestine it does so in liquid form. As it passes through the large intestine the water is reabsorbed, to maintain the body's hydration, producing dry stool. But if the colon empties quickly it has no time to complete this process, so the faeces remain mostly liquid. As losses mount on the trading floor, one observes anxious traders marching briskly to the toilets, the men's room starting to exude the fear and stench of a slaughterhouse.

THE STRESS RESPONSE TO A BEAR MARKET

Scott's amygdala, registering the severity of the situation, now switches on the big engines of the stress response by flooding his body with cortisol. Cortisol had already been released in small quantities before the news reports, giving Scott and the other traders something of a buzz; but it is now released, pulse after pulse, in such large quantities that it alters the character of the stress response, causing his body and brain to hunker down for a long siege. The effects of Scott's cortisol are now anything but pleasant. From here on, his attempts to remain cool and rational will encounter the same difficulties as a student trying to finish an exam in the middle of a fire drill.

The biology unfolds in the following way: the amygdala broadcasts a signal to the hypothalamus, a neighbouring brain region which controls the body's hormones. The hypothalamus tells the pituitary, a gland lying just below it, to secrete a chemical messenger into the blood, where it rushes about looking for receptors in which it fits. The messenger soon finds these in the rim of the adrenal glands, and instructs cells to manufacture cortisol. The cortisol, now pouring out of Scott's adrenal glands, carries a message to far-flung reaches of his body: the fight-or-flight is taking longer than expected, so to maintain the energy levels needed for this marathon struggle, shut down long-term functions of the body and marshal all available resources, mostly glucose, for immediate use. Adrenalin had initiated this process, but it is short-acting, so now cortisol takes over, maintaining high blood

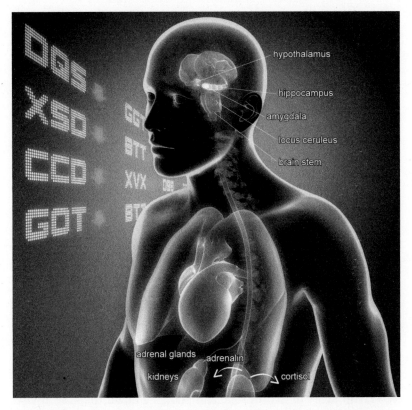

Fig. 10. The stress response. The initial and rapid phase of the stress response, called the fight-or-flight reaction, is triggered by the amygdala and locus ceruleus. The electrical signals of the fight-or-flight alarm travel down the spinal cord and out into the body, raising heart rate, breathing and blood pressure, and liberating adrenalin from the core of the adrenal glands. The more sustained phase of the stress response involves the hypothalamus, which, through a series of chemical signals carried in the blood, instructs the outer layer of the adrenal glands to produce cortisol. The cortisol then exerts widespread effects on both body and brain, instructing them to hunker down for a long siege by suppressing long-term functions such as digestion, reproduction, growth and immune activation.

pressure and an accelerated heart rate, and diverting energy away from digestion, reproduction, growth and energy storage.

Cortisol slows digestion by inhibiting digestive enzymes and shunting blood away from the stomach walls. It further inhibits the production and effects of growth hormone, stunting growth in young adults exposed to stress. Crucially, cortisol also reverses the body's anabolic processes. While an anabolic process builds up energy reserves, a catabolic process breaks them down for immediate use. Cortisol, as a catabolic steroid, blocks the effects of both testosterone and insulin; and it causes glycogen deposits to be broken down into glucose; fat cells into free fatty acids, an alternative energy source; and muscles into amino acids, which are then shunted to the liver to be converted into glucose. Cortisol has further effects in preparing us for a crisis: it suppresses the reproductive tract by inhibiting the synthesis of testosterone and sperm in men, and oestrogen and ovulation in women.

Finally, in case the crisis ends in injury, cortisol stands by as a powerful anti-inflammatory, one of the most effective known to medicine. In its role of preparing us for injury cortisol is aided by another powerful set of chemicals called endorphins, a type of opiate (responsible, some say, for the fabled runner's high), which are released in the body and brain during chronic stress as an analgesic, dulling our sense of pain. The effects of these natural painkillers are occasionally observed in battle when wounded soldiers fight on, unaware they have been injured.

Martin hears of Scott's loss, and glancing down the aisle advises him, 'This thing's a freight train; don't get in the way.' Scott wisely takes the advice, and over the course of the afternoon, as he tries to sell out the remains of his ill-conceived position, his cortisol levels continue to rise. He and Martin are having very different experiences, Martin thrilling to the volatility, Scott being crushed by it. In fact, across the floor, traders – depending on their physiology, training and exposure to credit markets – display varying physical responses to the volatility. Scott is a wreck, suffering an exaggerated stress response; Gwen is thriving on the flows, and is sustaining a mild fight-or-flight response, with moderate and invigorating levels of both adrenalin and cortisol, just as she used to in mid-tennis match; while Martin, benefiting from a toughened physiology and years of experience, has not needed much cortisol today, nor even much of the fight-or-flight response – his vagus nerve has merely released its brake on his heart and lungs, permitting the naturally powerful idle of his body to carry him through the afternoon without even breaking a sweat. Lucky him.

How does the vagal brake accomplish this miraculous feat? When you are in a relaxed state, reading a book say, your breathing and heart rate idle at a slow speed. But unlike a car, this resting heart rate is not your heart's default setting: its default setting is considerably faster, somewhere between a slow idle and full throttle. Your heart does not rev up to this natural setting because the vagus nerve applies its brake, slowing down both heart rate and breathing. If you are jarred out of this relaxed state by an emergency, the fight-or-flight nervous system takes over and raises your heart rate to a higher speed. But not for minor stressors. Between a resting heart rate and a fight-or-flight howl, there are intermediate levels of heart activation, and these are controlled by the vagus. In reaction to minor stressors the vagus can merely ease off its brake and allow the heart to accelerate on its own. This is a far more gentle and precise form of control over the heart and it is more efficient metabolically than launching into full fight-or-flight every time we confront a challenge. In fact we rely on these minute vagal adjustments to our heart throughout the day, and reserve the fight-or-flight acceleration for the times when trouble really looms. This is a marvellous trick. What a relief it is to let the vagus, like a trusty assistant, handle these minor hassles without causing us even a moment's concern. The physiological elite among us enjoy these benefits to an even greater degree: they have such good vagal tone that when faced by a vigorous challenge they do not need much cortisol, nor even much adrenalin, to handle it; they can merely release their vagal brake. Martin is lucky enough to belong to this physiological elite.

But not Scott. The crisis he faces today demands far more physiological resources than his idle can provide, so his body has initiated a powerful stress response. The tidal wave of stress hormones that now overwhelms him has been caused by more than the large amount of money he has lost; it has also been caused by the bewildering volatility of the market. Volatility means uncertainty, and uncertainty can have as large an effect on our bodies as actual harm, a fact of great importance in understanding stress in modern life.

In the early years of stress research, some scientists, such as Hans Selye, a Hungarian working at McGill University in the 1950s, believed the body mounted a defensive stress response largely to actual bodily harm, such as hunger, thirst, hypothermia, injury, low blood sugar, and so on. Others, some of them psychologists such as John Mason from Yale, noticed that the hypothalamus and the adrenal glands reacted more powerfully to the expectation of harm than to harm itself. Since then, researchers have found that three types of situation signal threat and elicit a massive physiological stress response – those characterised by novelty, uncertainty and uncontrollability.

Let us consider novelty first. When scientists exposed rats to a novel setting, by placing them in a new cage, the rats experienced an enhanced stress response, with elevated corticosterone (the rodent form of cortisol), even though nothing bad had happened and nothing in the environment presented an overt threat. This observation led scientists to suspect that the stress response was largely preparatory in nature: in novel situations we do not know what to expect, what can happen to us, so our adrenal glands pump out the stress hormones adrenalin and cortisol, which in turn sharpen attention and increase available glucose, just to be ready.

Uncertainty also powerfully affects cortisol secretion. In a series of intriguing experiments conducted in the 1970s, the endocrinologists John Hennessey and Seymour Levine found that an animal's stress response to a mild shock (nothing dangerous, just enough to make it withdraw its paw) depended more on the timing of the shock than on its magnitude. If a shock was delivered at regular or predictable intervals, or if it was announced by an audible tone, then after the experiment the animal might have normal or just slightly elevated cortisol levels. If, however, the timing of the shocks was altered so that they became less predictable, the animal's cortisol levels rose. As the timing of the shocks approached complete randomness, meaning they could not be predicted at all, cortisol levels reached a maximum. Animals received the same objective amount of shock in each experiment, yet experienced markedly different stress responses. Uncertainty about when the shock would come provoked more stress than the shock itself. Such a reaction is one we can all recognise, for it is the staple of horror films: we are more scared when uncertain about where the monster lurks than when it finally pops out and snarls at us. More seriously, it is also a pattern of stress that takes a heavy toll during times of war. During the Blitz in the Second World War, for example, the inhabitants of central London were exposed to daily, predictable bombing, while inhabitants of the city's outer suburbs were exposed to intermittent and unpredictable raids. It was in the suburbs that doctors found a higher incidence of gastric ulcers.

Uncontrollability has also been studied as a potent influence on stress levels. In a series of what are called 'yoked' experiments, two animals were given the same amount of shock, but one could push a bar lever to stop the shock for both of them. In other words, one had control, the other did not. At the end of the experiment the two animals had been exposed to identical amounts of shock (in that sense yoked), but the animal with no control displayed a more exaggerated stress response than the one with access to the bar lever. In later experiments it was found that the stress-reducing power of the bar remained even if it was unplugged and did nothing at all. Control, even the illusion of control, can mitigate the stress response, while loss of control in a threatening situation provokes the most terrifying stress response.

Novelty, uncertainty and uncontrollability – the three conditions are similar in that when subjected to them we have no downtime, but are in a constant state of preparedness. They are also the conditions in which traders spend a good part of their day. Do these features of their environment affect traders in the same way they do animals? The answer is, emphatically, yes. That is the conclusion I and my colleagues arrived at after our series of experiments with traders. One of these studies was discussed in the previous chapter, when I described the effects of testosterone on the traders' P&L. During that study, in addition to testosterone, I also collected cortisol from the traders, and gauged the uncertainty they faced by measuring the volatility of the market. The higher the volatility, we reasoned, the less certain traders would be of where market prices would trade in the coming days. What we found was that their cortisol levels rose substantially with the volatility of the market, demonstrating that their cortisol did indeed increase with uncertainty. In fact, so sensitive were traders' cortisol levels to volatility that they displayed a remarkably tight relationship with the prices of derivatives, the securities used to hedge volatility, a finding that raises the intriguing possibility that stress hormones form the physiological foundation of the derivatives market.

We also looked at the variability in their P&L, which is an indicator of how much control they have over their trading. This too showed that as the variability of their P&L rose, so too did their cortisol levels. The traders' hormone fluctuations, moreover, were extraordinarily large. In the normal course of events steroid hormones spike when we wake in the morning, this steroid surge acting much like a breakfast cup of coffee, and then decline over the course of the day. In this experiment we should have observed traders' cortisol levels dropping by about 50 per cent from morning to afternoon sampling times, but on volatile days they actually increased over the course of the day, some of them by an astonishing 500 per cent, levels normally seen only in clinical patients.

This afternoon, Scott finds himself trapped in a situation that is novel. He has never seen anything like this market – anything remotely like it – in his entire career. In fact, no one has. To find anything comparable, a crisis involving all credit markets and even threatening the solvency of governments themselves, you would have to go back to the Crash of 1929. Scott has furthermore never been so uncertain about the future course of events, an uncertainty he shares with other traders. Evidence of this collective uncertainty is found in the VIX, 'the Fear Index', which has risen from a sleepy 11 per cent in the summer to over 25 per cent today, and in the coming months will hit a terrified 80 per cent. Finally, Scott has lost record amounts of money, meaning by definition that he has lost control. The cumulative effect of losses and the novel, uncertain and uncontrollable conditions of the market is a massive upwelling of cortisol in Scott and other traders along Wall Street.

By four o'clock that afternoon Scott has been ordered by Ash to close out his positions, but he has not had much success, and finds it hard to concentrate. Part of the problem stems from a profound change that has taken place in his locus ceruleus. Earlier in the day, in response to the shocking analysts' reports, it had promoted a focused attention on the market and a heightened awareness of information relevant to predicting what mortgages would do next. But now, under a heavy load of stress, the pattern of neural firing in Scott's locus ceruleus alters, from short and frequent bursts to sustained firing. When this pattern takes over a person can no longer concentrate, but instead scans the environment, the reason being that when confronted with true novelty we no longer know what is relevant and what to focus on. Our scanning becomes hurried and indiscriminate, almost panicky. Too stressed to think clearly, his attention jumping from one thing to another, Scott sits out the rest of the day, powerless to trade profitably.

Over the next few days the news from the banking sector darkens even more, and traders realise with sinking spirits that the credit market will not recover any time soon. Scott unwinds the last of his mortgage trade only to find that his spread trade, stocks to bonds, is also losing massive amounts of money, as stocks follow mortgages into the abyss and Treasuries enter one of the fastest and most sustained rallies in history. By Friday Scott finds he has not only given back all his year's P&L, but has also lost an additional $9 million.

Scott had been looking forward to a weekend in the Hamptons with his girlfriend, taking in the late-autumn colours and the chill of sea air. But now he won't sleep much, or eat. His dreams of his own house on the beach have vanished like a gambler's lucky streak, and he wonders if he will even be able to afford a rental next summer. He spends most of the weekend on the phone to colleagues, reliving the week, collecting stories of other traders who have reassuringly also lost money. By Sunday his spirits have rallied somewhat. He may have given back his year, but, he reasons, his managers like him; the arb desk, despite losing $125 million last week, is still up $180 million on the year; and the bank as well has had a good year. He may no longer be in line for that $8 million bonus he expected, but he can piggyback off the arb desk's bonus pool and get, maybe, $1.5 million. After all, his girlfriend reassures him, the bank does not want to lose him to a competitor. Just to be safe, he starts showing up for work earlier than usual, wearing his best suits and ties, and having dinner with the salespeople he used to disdain. If you are not making money, you'd better at least have the sales force on your side.

But in the coming weeks Scott's optimism proves illusory. The markets have plunged into a financial crisis of historic proportions, and when in this angry state, they inflict maximum pain, searching out and dashing every hope. The Federal Reserve lowers interest rates once again, and will continue to do so in the months ahead, but these moves fail to ignite the expected rally in risky assets. The arbitrage desk, unable to get out of its positions, haemorrhages money at an alarming rate, not only giving back all its year's profits, but losing an additional $375 million. The bank is not in much better shape, with almost every department registering record losses. Over on the mortgage desk, Logan too has been sucked into the vortex now being called 'the Credit Crisis'. Despite his best efforts, client flows, all on the sell side, have kept him constantly long the mortgage market, and he has now lost more money this year than he has made in the past five.

Inevitably in these crises, just as night follows dusk, when nerves are close to snapping, rumours spread of impending layoffs, and uncertainty and uncontrollability reach new, debilitating levels. Traders, even former stars, feel vulnerable, and cannot count on job offers from other banks, let alone ones with a large signing bonus. There is even talk of the government shutting down all arb desks at the banks, even preventing flow traders from establishing arbitrage positions. As a result traders are starting to leave the banks for the hedge funds, where their risk appetite can still be fed. Desk managers now take to bullying juniors, hinting at changes to the desk, firing one or two people before the bank even announces layoffs. According to the primatologist Robert Sapolsky, dominant monkeys, when exposed to uncontrollable stressors, take to biting subordinates, an activity that horribly has the effect of lowering their cortisol levels, and managers, appearing to understand this ugly piece of physiology, offload their cortisol onto juniors, even ones who are performing well. In all the mess that goes into making a financial crisis, it is the uncertainty and uncontrollability created by middle managers that could most readily be minimised by upper management.

As December approaches and the days draw in, the stock market continues its plunge, and credit spreads, all of them, remain at historically wide levels. The high spirits of the bull market have now been thoroughly extinguished, and a wintry atmosphere descends on the trading floor. All across Wall Street, and overseas in the City of London and the financial centres of Tokyo, Shanghai, Frankfurt and Paris, the news is equally grim. Reports emerge of many black boxes being unplugged, the algorithms having as little success as humans in figuring out the financial anarchy; and hedge funds, even ones doing well, are seeing a flight of capital as investors shun risk. Scott and most other traders begin to suffer the toxic effects of a stress response that has gone on too long. It is at this point that cortisol has its most noxious impact on both brain and body, warping our thinking and damaging our bodies in ways that can kill us.

CHRONIC STRESS AND RISK-AVERSION

To understand the malevolent influence of stress in the financial world, we have to appreciate the difference between an acute exposure to stress hormones, i.e., moderate levels over a short period of time, and a chronic exposure, i.e., high levels over longer periods of time, for these two types of exposure have very different, and in most cases diametrically opposite, effects. Cortisol displays the same ∩-shaped dose–response curve we have encountered before, meaning that moderate levels have beneficial effects on cognitive and physical performance, while high levels impair them.

Moderate exposure to cortisol before the analysts' announcements enhanced the traders' vigilance, signal detection, metabolic preparedness and motor performance, and improved their mood almost to the point of euphoria. An acute reaction like this gave them a welcome edge. But the chronic exposure they have endured over the past month and a half is slowly poisoning them, wreaking havoc on their cardiovascular and immune systems, and in all likelihood impairing their ability to assess risk. The reason for this difference in effects is that the stress response evolved as a rapid, short-lived, and muscular retaliation; it was designed to switch on quickly, and to switch off after a short period of time. If it fails to do so, widespread medical problems ensue, largely because the stress response is metabolically expensive. The state of heightened readiness it promotes can be maintained over the long term only at the cost of breaking down many tissues in the body, much like burning furniture to keep a house warm.

Unfortunately, stress in the financial markets, and in society more generally, can indeed persist for long periods of time, because the ancient regions of the brain controlling the stress response – the amygdala, hypothalamus and brain stem – cannot distinguish clearly between a physical threat, which is usually brief (one way or the other), and a psychological or work-related one, which can endure for months, even years.

This latter situation is the one in which Scott now finds himself. Prolonged exposure to cortisol has begun to impair his ability to think and take risks almost to the point of making him useless as a trader. Part of the problem comes from a dramatic change that has taken place in the way his memory works. Cortisol affects memory by acting on dense receptor fields in the amygdala and in a neighbouring brain region called the hippocampus. These two brain regions act as a tag team in remembering stressful events. But they encode different aspects of memory: the amygdala the emotional significance of an event, the hippocampus the factual details.

This neural division of labour can be illustrated with the example of a child learning to ride a bicycle. After many false starts, the child finally kicks off, and lo and behold she is hurtling along the street unaided, a wonderful thrill. However, in her excitement she rides

208

straight through an intersection without looking, and narrowly escapes being hit by a car. She dissects the experience and stores pieces of it in widely dispersed parts of the brain, from the cerebral cortex down to the brain stem. The motor control behind the physical feat of riding may be locked up safe from the ravages of time in the cerebellum, a region of the brain that keeps functioning even if a patient suffers a complete amnesia. When referring to something you cannot possibly forget, people commonly say, 'It's like riding a bike.' The conceptual part of the learning experience, perhaps the point at which the girl figured out that the faster you ride the easier it is to remain on two wheels, may be stored away in her rational brain, the neo-cortex. The facts surrounding her first bike ride – the time of day, the location, the weather, who she was with, etc., in short her autobiographical memories – are stored in the hippocampus (although after a period of time they are moved from here to deep storage archives in the neo-cortex). And the fear caused by the near car accident may be stored in the amygdala. If this girl was to return to the same intersection a couple of years later, but with damage to her hippocampus, she might not recall the near accident, but her amygdala would fill her with dread, provoking a reaction with no more discernment than 'I'm scared, I don't like it here.' If on the other hand she returned with an intact hippocampus but a damaged amygdala she might remember every detail of the near accident but have no emotional reaction to the remembered event, the attitude of the hippocampus being, 'Just the facts, ma'am'.

Of these brain regions, and the types of memory they store, it is the amygdala and the hippocampus that are most affected by stress hormones. Through an extraordinary feat of chemical engineering, the same stress hormones that prepare our bodies to deal physically with a stressful challenge also instruct the amygdala and hippocampus to remember it, so we can avoid this or a similar risk next time. A mugging, a car accident, an encounter with a snake, news reports of 9/11 – tagged by cortisol for special storage, these events are captured for life as 'flashbulb memories'. Years later, even in old age, we seem to recall every detail surrounding them. Adrenalin, acting via the vagus nerve, assists cortisol in laying down these memories, and it has been suggested that administering beta-blockers, which inhibit the effects of adrenalin, just after a traumatic event may help prevent the creation of flashbulb memories, and may lower the risk of later panic attacks and post-traumatic stress disorder. At any rate, this week, when the mortgage market collapsed on Scott, wiping out his year, the events were seared into his memory.

Just as high levels of cortisol help us store traumatic events, so too do they later help retrieve memories of them. As cortisol levels rise, and our exposure to the hormone becomes chronic, we increasingly recall the events that were stored under its influence. Scott now finds he recollects mostly disturbing memories. He tends to dwell on nasty events – failing high-school calculus, a locker-room fight, losses during the dot.com crash – rather than pleasant ones, like meeting his girlfriend, vacations in Verbier, or trades he got right. Importantly, when assessing a trade, Scott now increasingly draws on negative precedents in determining the risks, and such a selective recall of things going wrong may promote an irrational risk-aversion.

Chronically elevated levels of cortisol have other effects on our thinking, besides those on memory, most importantly and disturbingly by changing the shape and size of various brain regions. Once again, the amygdala and the hippocampus, because they contain more cortisol receptors than other areas of the brain, are especially affected. If high levels of cortisol continue long enough they can kill neurons in the hippocampus, reducing its volume by up to 15 per cent, as they do in patients with Cushing's syndrome, a condition in which tumours in the adrenals or pituitary cause chronic overproduction of stress hormones. Fortunately the hippocampus is one of the few brain regions that can regrow neurons, so when the stress finally ends it can regenerate itself. Some neuroscientists, notably Bruce McEwen, believe this temporary loss of hippocampal volume serves to blunt the impact of stress on our brains. The hippocampus effectively hibernates through hard times.

Scott's hippocampus may shrink under the influence of cortisol, but his amygdala experiences the opposite effect. Neurons in the amygdala are fertilised by cortisol and undergo a rich arborisation (growth of branches), making Scott's thinking more emotional and less factual, and impairing his ability to engage in rational analysis. Some studies have even suggested that under conditions of extreme stress our pre-frontal cortex is effectively taken offline, impairing analytic thought and leaving our brains to run on stored reactions, largely emotional and impulsive ones.

Shell-shocked traders, under the influence of an overly active amygdala, become prey to rumour and imaginary patterns. In a recent study, two psychologists presented meaningless and random patterns to healthy participants, who appropriately found nothing of significance in them, and then to people exposed to an uncontrollable stressor, who did find patterns in the noise. Under stress we imagine patterns that do not exist. A striking real-life example of this phenomenon is reported by Paul Fussell in his astonishing book *The Great War and Modern Memory*. Troops living in the trenches during the First World War, under the most unimaginable conditions of fear and uncertainty, were deprived of reliable information about the war because the official army newspaper contained little but inaccurate propaganda. In the absence of reliable information, and in desperate need of it, troops fell prey to rumour in a manner not seen since the Middle Ages – rumours of wraithlike spies conversing with frontline troops before disappearing into the mist; of angels in the sky over the Somme; of a factory behind enemy lines called the Destructor where bodies of Allied soldiers were rendered for their fats; of tribes of feral deserters living in no-man's land, preying on injured soldiers. Traders during a financial crisis suffer from an equally wretched vulnerability to rumour and suspected conspiracy. Every bank, individually or collectively, at one time or another is going under; hedge funds, huge ones of course, conspiring to push down the markets; the Chinese dumping Treasuries; the UK defaulting on its sovereign debt; broker suicides. Each rumoured catastrophe is now given as much credence, and has as much effect on markets, as hard economic data.

Cortisol's lethal effects on the brain are compounded by another chemical produced during stress, one produced in the amygdala called CRH (short for corticotropin-releasing hormone). CRH in the brain instils anxiety and what is called 'anticipatory angst', a general fear of the world leading to timid behaviour. Together with cortisol, it also suppresses the production of testosterone, the invigorating hormone that powered so much of Scott's confidence, exploratory behaviour and risk-taking during the bull market. He now scares easily. He develops a selective attention to sad and depressing facts; news comes freighted with ill portent; and he seems to find danger everywhere, even where it does not exist. This paranoia colours his every experience; and when riding home in the taxi at night Scott finds that even his beloved New York City, once sparkling with opportunity and excitement, has lately taken on a menacing silhouette. As a result of chronic stress he, like most of his colleagues, becomes irrationally risk-averse.

By mid-December, the financial industry has endured a month and a half of endless volatility and non-stop losses. The run-up to Christmas is normally one of the most optimistic and playful times of the year, with the holidays and skiing vacations to look forward to, followed by bonus payments in the New Year. But such gaiety as had survived the crash has now been crushed by layoffs, brutally announced just before Christmas, involving almost 15 per cent of the sales and trading staff. Few people will get any bonus at all; and the lucky ones, like Martin and Gwen, who do get a small one, harbour a deep resentment because this year they have made record profits and helped to keep the bank afloat, while traders like Stefan, paid over $25 million last year, have helped blow up the bank and with it their, Martin and Gwen's, bonuses. Scott will get nothing at all, and does not know how long he will be kept on. Layoffs have been similarly announced all along Wall Street and in the City of London. Many firms, facing bankruptcy, have closed their doors. One by one, the lights are going out all across the financial world.

With their jobs on the line, traders like Scott desperately need to make money, but find themselves oddly unable to initiate a trade, even one that looks attractive, being held back from the phones as if by a force field. They have become, as they say in the business, 'gun shy'. A reduced risk-taking among traders would be a welcome change under normal conditions, but during a crash it poses a threat to the stability of the financial system. Economists assume economic agents act rationally, and thus respond to price signals such as interest rates, the price of money. In the event of a market crash, so the thinking goes, central banks need only lower interest rates to stimulate the buying of risky assets, which now offer relatively more attractive returns compared to the low interest rates on Treasury bonds. But central banks have met with very limited success in arresting the downward momentum of a collapsing market. One possible reason for this failure could be that the chronically high levels of cortisol among the banking community have powerful cognitive effects. Steroids at levels commonly seen among highly stressed individuals may make traders irrationally risk-averse and even price insensitive. Compared to the Gothic fears now vexing traders to nightmare, lowering interest rates by 1 or 2 per cent has a trivial impact. Central bankers and policy-makers, when considering their response to a financial crisis, have to understand that during a severe bear market the banking and investment community may rapidly develop into a clinical population.

Of the conditions affecting traders, a particularly unfortunate one is known as 'learned helplessness', a state in which a person loses all faith in his or her ability to control their own fate. It has been found that animals exposed repeatedly to uncontrollable stressors may pathetically fail to leave the cage in which this experiment was conducted if the door was left open. Traders, after weeks and months of losses and volatility, may similarly give up, slumping in their chairs and failing to respond to profit opportunities they would only recently have leapt on. In fact there is some evidence suggesting that people like traders might be especially prone to this sort of collapse. Banks and hedge funds commonly select traders for their tough, risk-taking, optimistic attitude. Optimism is generally a valuable trait in a person, especially a trader, for it leads them to welcome risk, and to thrive on it. But not always. Not if they are exposed to long-lasting and unpredictable stressors. Research has suggested that optimistic people, those who are used to things working out, may not handle recurrent failure very well, and may end up with an impaired immune system and increased illness. Bankers, so well suited to the bull market, may be constitutionally ill prepared to handle bear markets.

A telling sign of the onset of learned helplessness is the subsiding of anger on the trading floor, anger being in fact a healthy sign that someone fully expects to be in control. During a crisis, when swearing dies down, fewer phones are smashed, and anger is replaced by resignation, withdrawal and depression, chances are traders have succumbed to learned helplessness. Once stress in the financial world has reached this pathological state, governments must step in, as they did in 2008–09, and do the job that traders can no longer perform – buy risky assets, reduce credit risk, lead the traders, now reduced to a shellshocked state, out of the slough of despond.

STRESS-RELATED DISEASE IN THE FINANCIAL INDUSTRY

Prolonged and severe stress endangers more than the financial system: it poses a serious threat to the personal health of people working in the financial industry, and indeed in all the industries affected by troubles in the banking sector. In the workplace the difference between acute and chronic effects is most worryingly apparent. A prolonged stress response, by shutting down so many long-term functions of the body, impairs its ability to maintain itself. Blood has been shunted away from the digestive tract, so people become more susceptible to gastric ulcers. The immune system, thrown into overdrive during the early stages of the stress response, has after chronic exposure to cortisol been suppressed (possibly because it draws too much energy), so people find themselves constantly battling upper respiratory diseases, like colds and 'flus, and other recurrent viruses, like herpes. Growth hormone and its effects have been suppressed, as have the reproductive tract and the production of testosterone.

This last effect, in addition to tensed muscles which prevent blood flow into what are called the cavernous cylinders (corpora cavernosa) of the penis, causes bankers like Scott, sexually insatiable during the bull market, to have difficulty maintaining an erection, even mustering any interest in sex, testosterone being the chemical inducement for erotic thoughts. Chronic stress, largely through cortisol's interaction with the dopamine system, can also make people more susceptible to drug addiction. And all these effects are magnified by the fact that elevated cortisol levels reduce sleep time, especially REM sleep, thereby depriving people of the downtime needed for mental and physical health. Steroids may orchestrate a symphony of physiological effects, but as time passes the music turns into a cacophony.

Perhaps the most harmful effect of prolonged stress is the chronically raised heart rate and blood pressure, a condition known as hypertension. The unceasing pressure on arteries that comes with hypertension can cause small tears in arterial walls, tears which then attract healing agents called macrophages or, more commonly, white blood cells. Mounds of these sticky clotting agents grow over the arterial injuries, and subsequently trap passing molecules, like fats and cholesterol. Larger and larger plaques form, which can become calcified, a condition known as atherosclerosis, or hardening of the arteries. As the plaques grow and block arteries, they decrease blood flow to the heart itself, causing myocardial ischemia, or angina, a recurring pain in the chest. If the plaques become large enough they may break off, producing a thrombus, or clot, which then travels downstream to smaller and smaller arteries, and ends up blocking an artery to the heart, causing a heart attack, or an artery to the brain, causing a stroke.

As the economic crisis deepens, cortisol's catabolic effects add to the problems created by high blood pressure. Insulin, which normally withdraws glucose from our blood for storage in cells, has been inhibited for months now, so high levels of glucose and low-density lipoproteins, the so-called bad cholesterol, course through traders' arteries. Muscles as well get broken down for their nutrients, and the resulting amino acids and glucose circulate needlessly in the blood, looking for an outlet in demanding physical struggle. Our stress response is designed to fuel a muscular effort, yet the stress most of

215

us now face is largely psychological and social, and we endure it sitting in a chair. The unused glucose ends up being deposited around the waist as fat, the type of fat deposit posing the greatest risk for heart disease. At the extreme, stressed individuals, with elevated glucose and inhibited insulin, can become susceptible to abdominal obesity and type 2 diabetes. Patients suffering from Cushing's syndrome epitomise the change in body shape, having atrophied arm and leg muscles and fat build-up on the torso, neck and face, making them appear much like an apple on toothpicks. A year into the financial crisis, the testosterone-ripped iron men of the bull market start to look decidedly puffy.

Heart disease caused by uncertainty and uncontrollability in the workplace has been amply documented. In a pioneering study of job stress, *Healthy Work*, Robert Karasek and Tores Theorell found that workers facing the highest levels of workload coupled with uncontrollability in their jobs suffered higher rates of hypertension, elevated cholesterol and heart disease, all signs of chronically elevated stress hormones. Similarly, in Britain, a series of studies called the Whitehall Studies looked at stress among civil servants, most notably in departments undergoing privatisation. The authors found that the employees most exposed to job insecurity suffered higher levels of cholesterol, higher rates of weight gain, and an increased incidence of stroke. Finally, epidemiological evidence exists of widespread damage that economic recessions wreak on the health of workers. A study conducted in Sweden with 40,000 people over a sixteen-year period found that health was strongly correlated with the business cycle, with mortality from cardiovascular disease, cancer and suicide all increasing during recessions.

Data on the financial industry is sparse, but private health insurance companies throughout the US and the UK reported a surge in claims for peptic ulcers, stress and depression after the credit crisis began in the autumn of 2007. In July 2008, for example, British United Provident Association Ltd, the UK's largest private health insurer, reported that the number of employees from financial institutions seeking treatment for stress and depression had risen 47 per cent from a year earlier. The World Health Organisation also warned about the rise of mental-health problems and suicide with the advent of the credit crisis and the subsequent recession. Recently, a few years after the event, we are starting to get hard epidemiological data on the fallout from the Credit Crisis. One study has found that during 2007–09 there was a spike in the rate of heart attacks in London, and this occurred against a backdrop of a decreasing incidence of heart attacks in the rest of the UK. The authors estimate that this surge in heart attacks in London led to an additional 2,000 deaths, and resulted, they suggest, from the impact of the Credit Crisis on the financial district. A market crash may thus produce not only an economic disaster but also a medical one.

In the myriad ways described here, the stress response, as it builds and ramifies over the course of weeks and months, worsens the credit crisis. The bodily response initiated to handle the stress feeds back on the brain, causing anxiety, fear and a tendency to see danger everywhere. By so doing, this steroid feedback loop, in which market losses and volatility lead to risk-aversion and to a further sell-off in the market, can exaggerate a bear market and turn it into a crash. Body–brain interactions may thus shift risk preferences systematically across the business cycle, destabilising it. Economists and central bankers, such as Alan Greenspan, refer to an irrational pessimism upsetting the markets, just as John Maynard Keynes once spoke of the dimming of animal spirits. With the development of modern neuroscience and endocrinology we can begin to provide a scientific explanation for these colourful phrases: cortisol is the molecule of irrational pessimism.

PART IV

Resilience

EIGHT

Toughness

CAN WE CONTROL OUR STRESS RESPONSE?

Watching cortisol at work, as we have on our visit to Wall Street, enables us to see what many endocrinologists have long recognised: evolution has equipped us with a stress response that can be fatally dysfunctional in modern society. As it lingers and becomes chronic, as it so easily does with work-related or social problems, the stress response morphs from life saver to killer. It may have been engineered to carry us clear of immediate danger, but like the fire brigade, the stress response may save our house from an emergency only to destroy it with water damage. Indeed, chronic stress may be responsible for many of the most deadly and intractable problems faced by modern medicine – hypertension, heart disease, type 2 diabetes, immune disorders and depression.

Given what is at stake, for both personal health and the stability of the financial system, we have to ask: can we turn off the cortisol? Can we control its toxic body–brain feedback loop? Sadly the answer is: only with very great difficulty. Our conscious and rational selves have very little control over sub-cortical parts of the brain such as the amygdala, the hypothalamus and the brain stem. The trouble, as Joe LeDoux explains, is that we have a forest of axons (the fibres sending messages from a neuron) running up from our brain stem and our limbic system (the emotional brain) to our neo-cortex, ensuring that our rational efforts are regularly influenced by sub-cortical signals; but we have many fewer axons extending down into these primitive brain regions, so proportionally less conscious influence over them. Anyone who has experienced ungrounded panic attacks, or been in love with the wrong person, knows that efforts to consciously change our feelings are doomed to an endless cycle of repetition and failure.

An illuminating demonstration of the almost complete divorce between the conscious and unconscious expression of stress can be found in what is called an open field experiment conducted with rodents. When researchers put a rodent in an open field – a dangerous place, given its easy exposure to predators – it displayed classic symptoms of an animal stress response: freezing into immobility, defecation and elevated corticosterone (the rodent version of cortisol). However, if the researchers repeated this procedure for several days, the rodent gradually habituated to the experience. Nothing bad had happened yet, so the behavioural side of the stress response abated; the rodent stopped freezing and defecating. Interestingly, though, its corticosterone levels remained stubbornly elevated. The rodent was no longer consciously registering stress, yet its hormones were.

Now ask yourself: which of these two responses, the behavioural or the physiological, is more appropriate to the situation? Well, a rodent should not be in an open field – an objectively dangerous place for it – so it should indeed be stressed. And, remarkably, its adrenal glands know this even if its conscious brain does not.

We found much the same thing with the traders we were studying. In the chapter on gut feelings I described an experiment in which we sampled cortisol from traders and asked them through a questionnaire how stressed they were by their P&L or by the markets. Their answers turned out to have little if anything to do with losing money, large swings in their P&L or high volatility in the market; yet their cortisol levels did faithfully track these stressors. Our findings illustrated just how disconnected the conscious and unconscious stress responses can be, how people often invent stories to accompany their behaviour. By means of these stories we may even persuade ourselves we are not stressed, or talk ourselves into feeling better about our plight. Yet if our objective situation remains novel, uncertain or uncontrollable, our physiology will remain on high alert, and in time our health will suffer. The hypothalamus and the adrenal glands appear to respond more to objective cues than to an encouraging chat. Their pathologies may not be amenable to a talking cure.

Despite this seemingly bleak conclusion, research into the physiology of the stress response holds out more promise than discouragement. First, by allowing us to see that stress is largely a physiological preparation for physical action, this research raises the possibility of training our physiology so that we develop a greater mental and physical stamina, toughening us against the fatigue, anxiety and psychiatric disorders that follow from chronic stress. Such a possibility may sound futuristic, yet there is a field of science that has already made remarkable breakthroughs in designing just such toughening regimes, and that is sports science. Second, by allowing us to see that stress emerges from objective circumstances, the research raises the possibility of changing these circumstances, changes which could then filter through to our mental state and our physical health. Let us consider in turn these two approaches to mitigating chronic stress: physiological toughening and objective changes in the workplace.

THE TOUGHENED INDIVIDUAL

Physical toughness is today relatively well understood. Sports scientists have made great advances in their understanding of strength, posture, coordination and endurance. Mental toughness, by comparison, has received far less attention, and remains accordingly less well understood. This is unfortunate. Work today relies less on physical effort and more and more on mental effort, and with this change has come a greater number of work days lost to anxiety, mental fatigue, stress and depression.

The research that does exist on mental toughness, coming from physiology, neuroscience and sports medicine, nonetheless offers some tantalising suggestions. To begin with, mental toughness involves a particular attitude to novel events: a toughened individual welcomes novelty as a challenge, sees in it an opportunity for gain; an untoughened individual dreads it as a threat and sees in it nothing but

223

potential harm. What is intriguing about the research into toughness is the finding that to each of these attitudes – viewing novelty as a challenge or as a threat – there corresponds a distinctive physiological state.

Medical researchers and sports scientists have studied the differences between these physiological states in order to answer a number of medically important questions. What neuro-chemical profile characterises a person who can cope effectively even when scared? Why can some people maintain low levels of anxiety even when faced by uncontrollable stressors? And what balance of hormones and neuro-modulators enables some people to remain motivated even in an environment lacking rewards (perhaps an environment like a financial crisis)? Medical researchers believe that finding answers to these questions will help them mitigate the effects of chronic stress, diagnose and prevent depression, and understand and treat conditions such as post-traumatic stress disorder, a debilitating syndrome affecting war veterans and people who have suffered personal trauma in which they vividly and uncontrollably relive their terrors.

Some scientists, recognising that mental toughness corresponds to a physiological profile, have gone a step further and asked, can this toughness be trained? Can purely physical training regimes translate into emotional stability, mental endurance and improved cognitive performance? Scientists who think the answer is 'yes' have built their research upon a curious finding – that resilience to stress comes from experiencing stress.

This idea originated in a lab at Rockefeller run by a psychologist named Neal Miller. Miller was one of the fathers of what is called behavioural medicine, the idea that behavioural therapy can rewire our brain and rebuild our body just as thoroughly as many medications. He and his lab also conducted some pioneering experiments in the physiology of stress. It was in Miller's lab that two of his students, Bruce McEwen and Jay Weiss, discovered cortisol receptors in the brain, and in the process described the hormone feedback loops between brain and body which, I have suggested, may be influential in the financial markets. Miller also made some of the first discoveries about toughening. In particular, he and Weiss found that when rats were exposed to chronic (in other words unrelenting) stress they came to suffer both physical illness and learned helplessness, and these were the result of depleted noradrenalin levels in their brains. However, if the rats were exposed to acute (in other words short-lived) stress, even if it was repeated over and over, they emerged with a hardier physiology and an increased immunity to the damaging effects of further stressors. These findings, initially surprising, enabled scientists to see that the process of mental toughening bears similarities to that of physical toughening.

Sports scientists know, for instance, that to build lean-muscle mass and expand aerobic capacity athletes must endure a training process that shocks their muscles and taxes their cardiovascular systems, to the point of inflicting mild damage to tissues, and then punctuate this process with periods of rest and recovery. Stress, recovery, stress, recovery – when calibrated to exhaust an athlete's resources, but only just, and then replenish them, the process can expand the productive capacity of a broad range of cells in the athlete's body. When coaches time this training regime just right they can tune their athletes so that they arrive on the day of competition with the optimal amount of glucose, haemoglobin, adrenalin, cortisol and testosterone coursing through their arteries. What the scientists studying toughening have found is that a somewhat similar process of challenge and psychological loading followed by recovery can tune our brain and nervous system so that we too approach stressors with resilience and an optimal mix of hormones, neuromodulators, and nervous-system activation.

What exactly does this resilience consist of? What is physiological toughness? And how do we achieve it? To describe the state of toughness and how the process of toughening works we should consider in turn each ingredient in its physiological cocktail: catabolic hormones; anabolic hormones; amines, the class of chemical which includes adrenalin, noradrenalin and dopamine; and the vagus nerve.

Catabolic hormones. Sports scientists and medical doctors know that our catabolic mechanisms, such as cortisol-producing cells, must be kept on a short leash. To repeat, a catabolic hormone is one that breaks down energy stores, such as muscle, for immediate use. Cortisol, as mentioned, is thus crucial in supplying us with energy when we mount an all-out physical or mental effort, but it is in many ways too powerful, and should be administered sparingly. By breaking down muscles and converting them into immediately usable forms of energy, cortisol in effect strip-mines our body for nutrients. If it is not turned off quickly, after a few days, a couple of weeks at the most, our body begins to disintegrate under its caustic influence. We come to suffer, as surveyed in the previous chapter, a broad range of physical ailments, as well as anxiety and a tendency to view events as threats rather than challenges.

We need cortisol to deliver metabolic support when we are challenged, but we must prevent it from turning into some doomsday-like defence mechanism, saving us for the moment but guaranteeing long-term annihilation. Its production and release should therefore occur sporadically, much as it would in a well-timed training regime, and be followed by a recovery period. A regular ebb and flow of catabolic hormones promotes health, but a continual flow kills.

Anabolic hormones. It is during our downtime, when catabolism is turned off, that anabolic hormones step in and rebuild our depleted energy stores so that we have fuel to draw on next time we are called into action. These anabolic hormones include testosterone and growth hormone, which together convert amino acids into muscle and calcium into bone; insulin, which removes excess glucose from the blood and deposits it in the liver; and a chemical called insulin-like growth factor (IGF), which rejuvenates cells throughout the body and brain. A healthy person, and to a greater extent a tempered athlete, will have a high ratio of anabolic to catabolic hormones, a ratio known as the growth index. A high growth index indicates a robust capacity to rebuild one's body after a period of destruction, a condition that Bruce McEwen, together with Elissa Epel and Jeannette Ickovics, has termed 'thriving'.

Without downtime our growth index declines, with the result that even a strenuous athletic training regime will fail to deliver results, the athletes finding to their frustration that their performance has gone stale. Older people may suffer a more serious decline in their growth index because they may stop producing testosterone and growth hormone altogether, while producing ever increasing amounts of cortisol; as a result they come to suffer what is called a 'failure to thrive', their high cortisol levels draining them of muscle and vitality. The simple ratio of testosterone to cortisol, easily assessed through either a saliva or a blood sample, can serve as a sensitive measure of our immunity to daily stress and our state of preparedness for competition. However, McEwen and colleagues recommend a slightly more complex measure, an index of the body-wide strain we experience when stressed. McEwen includes in this index blood pressure, body-mass index, hip-to-waist ratio, cholesterol levels, blood glucose levels, and noradrenalin and cortisol levels as sampled from urine. He has found that this index, more than any one of its components on its own, is a reliable predictor of future health.

Amines. A rhythmic alteration in the body between stress and rest, between the actions of anabolic and catabolic hormones, builds endurance. Some research has tentatively but tantalisingly suggested that such a regime may also expand the productive capacity of our amine-producing cells. These cells manufacture dopamine, noradrenalin and adrenalin, as well as many other chemicals, such as serotonin, the main target of antidepressant drugs such as Prozac. Amines switch on quickly, focus our attention, release glucose, and promote the full flight-or-fight response, just as cortisol does; but since the half life of amines in the blood is only a few minutes, they switch off as soon as the stress ends. According to Richard Dienstbier, one of the first scientists to work on toughness, a toughened individual is one who enjoys a powerful and immediate amine reaction when challenged, so he or she does not need to draw on the longer-acting and more potent cortisol response.

In a toughened individual, amine levels are lower at rest, rise more strongly when stressed, and shut off quickly. Since this person's physiology is capable of handling the stressors being thrown at them, their homeostasis is not thrown out of balance, so they handle the stress without emotional distress. Physiological coping and emotional distress seem to be alternatives – if your body is coping, why get upset? As we saw when discussing homeostasis, emotions erupt, urging us to try alternative behaviours, when our body cannot handle a crisis on autopilot. The research into toughness has suggested that our brain silently compares the demands being made on us against the resources we can draw on (taking into account our training and skill). If our resources are sufficient we view the event as a challenge and relish it; if not we see it as a threat and shrink from it.

Our wonderful little amine-producing factories can, however, be overworked. Should our amine-producing cells be denied a rest period they become depleted; and then we are left to handle daily challenges without their energising influence. Depleted amines lead to a range of psychiatric and clinical disorders. Dwindling dopamine reserves, for example, can leave us lacking motivation. One of the symptoms of depression is a condition known as anhedonia, the failure to experience pleasure in life, any pleasure, even from favourite foods or activities; and anhedonia occurs with depleted dopamine cells. Similarly, depleted noradrenalin cells can leave us chronically lacking in arousal and enthusiasm. Worse, it can lead to the learned helplessness that Scott and Logan suffered. Such a condition can be brought on if we are exposed to unrelenting stress, such as combat, or divorce, or a two-year-long credit crisis, leaving us to brood over problems night and day, the troubling thoughts keeping the locus ceruleus on full alert, depriving it of any downtime, until eventually it is depleted of its precious noradrenalin. It is common among depressed people to have depleted noradrenalin and dopamine on the one hand, and chronically high levels of cortisol on the other.

What is remarkable about the research into toughening is the discovery that these amine-producing cells, like muscles, not only need a recovery period to rebuild their inventories, but can be trained to increase their productive capacity. The greater this capacity, the less likely they are to become depleted during stress, the more likely we are to view events as challenges, and the less likely we are to draw on the more damaging cortisol response. A strong first response by amines is the sign of someone who is coping; a strong cortisol response, someone who is not.

The picture of a toughened physiology that emerges from research on amines and hormones thus looks something like this: a toughened individual has a high ratio of anabolic to catabolic hormones. When faced by a challenge, the toughened individual experiences rapid and large increases in the amines, both in the brain and in the body, followed by moderate increases in cortisol. A toughened person, counter-intuitively, has a stronger initial stress response than an untoughened person, but he or she masters the situation, permitting the cortisol to abate, while the untoughened person mounts a weak arousal but the cortisol lingers, causing catabolic damage. Importantly, toughened people endure a sustained challenge without depleting the amines in their brain or succumbing to learned helplessness. Such a profile provides a person with all the cognitive and metabolic benefits of the amines while suppressing the damage done by a chronic exposure to cortisol. Such a profile is found, for example, in elite athletes.

It is also found in good traders. In one of the studies we conducted, I and my colleague found that the most experienced and profitable traders displayed extremely high and volatile steroid hormone levels, both testosterone and cortisol. The finding was initially puzzling, since we, like most people, expected veteran traders to be hardened and unemotional. And in fact they were, displaying little emotion through wins and losses. Nonetheless, behind their poker faces there roared an endocrine system on fire. In light of the research on toughened athletes, the finding makes perfect sense. Like Olympic athletes, these guys could call on their hormones when needed, and when the challenge had passed, the hormone levels rapidly returned to baseline before they could do any damage. Amateur traders, like amateur athletes, display the opposite profile: they have chronically raised cortisol levels, leaving them in a state of anxiety in which stressors come to be viewed as threats to be feared rather than challenges to be tackled.

The vagus nerve. We must now weave into the fabric of a toughened physiology the role played by the rest-and-digest nervous system, and the vagus nerve in particular. The fight-or-flight nervous system prepares us for vigorous, even violent action, but the rest-and-digest system takes over once the action has finished. It is, to extend Shakespeare's apt phrase, 'great nature's second course', and together with anabolic hormones it knits up the body's 'ravell'd sleeve of care'.

The calming effects of the vagus have led Stephen Porges to view it as a highly evolved and efficient tool for conserving energy. Porges has traced the history of the vagus nerve and found that as it evolved from its simplest state in reptiles to its more complex form in mammals, it played a role in three successive stress responses: freezing, fight-or-flight, and social engagement. Porges's history is worth recounting because it suggests that the advanced vagus nerve we have inherited today may provide us with one of our most valuable resources for minimising the ravages of stress.

In reptiles, the vagus orchestrates a primitive reaction to threat – freezing into a motionless state. Reptiles freeze in order to conserve their limited energy and avoid detection. This freeze response was subsequently passed on to mammals, in which it proved useful as a way of feigning death when a threat loomed. Something like the freeze response is also activated in mammals living or feeding in water, such as seals, in order to slow heart rate and metabolism and conserve oxygen when diving to great depths. The vagal freezing response lingers to this day in most mammals, and can show up under circumstances of extreme danger. When escape from a predator is deemed impossible, a mammal can call on this ancient reaction, and its physiological systems will more or less shut down. Movement ceases, breathing slows, heart rate drops, sensitivity to pain decreases – pathetically, the animal may provoke cardiac arrest and die from its own reaction to threat rather than from the predator itself. Something like this phenomenon has been observed in wild rodents trapped in water they fear they cannot escape. These poor animals, realising the futility of struggle, frequently opt for freezing and cardiac arrest, some of them even diving into the depths, drowning themselves. Presumably they do so to trigger a dive-induced freezing, leading to a rapid and painless death.

We humans retain this prehistoric freeze reaction. You can verify this fact by dipping your face into cold water (only your face will do), for this action triggers the dive reflex, which slows the heart rate and breathing, and quite possibly gives you a shot of natural painkillers. When highly stressed, people often splash cold water on their faces to engage this reaction, even if they have no knowledge of the physiology involved. Vagal freezing may also show up, some scientists have suggested, in cases of sudden death caused by the receipt of shocking news, and even in documented cases of voodoo death, an inexplicable death brought on, presumably, by the complete conviction held by a person that the curse just placed on them will prove effective.

In the next stage in the evolution of the vagus, this nerve came to cooperate with the fight-or-flight response. During fight-or-flight the vagus removes its slowing influence, what Porges calls the vagal brake, from visceral organs so that fight-or-flight can take over, just as it did when Martin and Gwen heard the Fed announcement and launched into action.

It is in the third stage of its evolution that the vagus nerve reaches its most sophisticated and encouraging form – as a tool of approach and conciliation. Porges sees the human vagus as a system of social engagement, an evolutionarily advanced and metabolically efficient alternative to fight-or-flight that promotes diplomacy instead of confrontation. Speaking in a calm and reassuring voice, making eye contact, displaying facial expressions that broadcast cooperation rather than confrontation, all these help avoid a metabolically expensive and potentially damaging fight; and crucially they calm our visceral arousal. The vagus nerve, it could be said, forms the diplomatic core of the body.

Today we harbour within our bodies all three vagal reactions. Each of these may be initiated when we are caught in an escalating confrontation, beginning with the most recently evolved one, and then proceeding to the older ones. Our first reaction to a challenge is accordingly social engagement, in other words, talking, making eye contact, calming down the situation. If this diplomacy fails then we reluctantly fall back on the older fight-or-flight reaction. Should even this fail and neither victory nor escape from the threat proves possible, then we may lapse into the ancient reptilian state of freezing or giving up or feigning death or in extreme and very rare cases dying a voodoo death. In the course of this confrontation we have regressed millions of years in evolutionary time.

Our highly advanced vagus nerve permits us to subtly adjust our stress response to the demands made on us. By doing so, it conserves energy. When confronted by daily stressors the vagus merely eases off its brake, permitting our body's idle to speed up just enough so that we can handle mundane challenges without bringing online the more costly fight-or-flight or cortisol systems. This may be one reason that people who view an unexpected event as a challenge have been found to display efficient cardiac output coupled with low blood pressure in their peripheral arteries; while people who view the event as a threat have inefficient cardiac output and high blood pressure. Indeed, people who lack a well-functioning vagus, who suffer from what Porges calls poor vagal tone, tend to overreact to mild stressors, and instead of calibrating a subtle response to these mundane situations they launch into a full fight-or-flight confrontation. Uproar is their only music. The lack of good vagal tone drains a person's energy and ultimately their health. Porges has found that children with low vagal tone display more behavioural problems later in life.

On the other hand, truly toughened individuals, the physiological elite, such as world-class athletes, may have been gifted with bodies and brains so primed for maximal effort that they appear almost otherworldly in their ability to cruise through a gruelling physical contest with nothing more than a slight release of their vagal brake. Can they really be performing that well just on their body's idle speed?

Vagal tone can be measured through a person's heart-rate variability. When you breathe in, your heart rate speeds up; when you breathe out, it slows down. This speeding up and slowing down is governed by the vagus nerve. People with good vagal tone will have heart rates that are highly variable. This variability is a good thing – the slowing down giving our hearts what amounts to a mini-rest every breath – and has been correlated with a number of markers of health. (It should not be confused with palpitations or heart arrhythmias – these are irregular heartbeats, but not ones that occur systematically across every breath.) On the other hand, people with poor vagal tone, or people who are stressed, will show little or no heart-rate variability, their hearts pumping at a constant rate. This lack of heart-rate variability is a risk factor for hypertension and future heart disease. Heart-rate variability can be monitored easily with a small device, sold commercially, worn on your chest or clipped to a finger.

We can therefore add good vagal tone and a high level of heart-rate variability to our profile of a toughened individual. In fact, one study of soldiers has found that a high level of heart-rate variability generally co-existed with a high ratio of anabolic to catabolic hormones.

In short, a toughened individual, one who views novel events as a challenge to embrace, draws on very different physiological systems from an individual who views them as a threat to avoid. Their different responses could be described as offence versus defence. The former is energising and enjoyable, leading to the coveted state of flow; the latter is draining and unpleasant, leading to a fear of the world.

Incidentally, the dramatically different physiological experiences of challenge and threat may be an underappreciated influence in social confrontations such as legal disputes and political battles. Take the example of a group of villagers fighting a developer in order to preserve their green spaces, or the managers of a company facing a hostile takeover. These people are confronted by developers and take-over sharks who live for such battles, revel in them. But not the defenders. For them the battle is a desperate and unpleasant experience, soaked in stress hormones, and it leaves nasty memories that may dissuade them from getting involved in future confrontations. It is often said that in war the best defence is a good offence; and perhaps in politics, business and sport, there is physiological support for this bellicose piece of advice.

THE SCIENCE OF GRACE UNDER PRESSURE

Can we toughen our physiology? Inevitably, a good deal of our toughness comes from the genes we inherit. Certain genes, for example, make some people more immune to the effects of stress and the stress hormones. But some scientists have found that developmental influences affect the way a person handles stress later in life. They have found that acute stress, in other words short-lived and moderate stressors, early in life can toughen an animal for adulthood. Young rats that are handled by humans will develop larger adrenal glands, but nonetheless as adults will show a more muted stress response to threats. They also tend to live longer, one study finding a life expectancy 18 per cent longer than that of non-stressed rats. The acute stressors must, however, be acute and moderate, for the same research showed that major stressors early in life, such as maternal separation, foster an anxious adult ill-prepared to deal with the slings and arrows of normal life events.

The training or toughening effects of acute intermittent stress can also be observed in adult rats. These stressors may include handling by humans, running on an exercise wheel, mild shock, even having their amines depleted by drugs. It does not seem to matter that much what the stressor is. The stress response is a general, bodywide reaction, so any stressor can exercise it. Each of these stressors, if brief and repeated, could toughen the rats. Call it the school of hard knocks.

What stressors could contribute to toughening in humans? Research on toughening regimes is still in its infancy, but a few types of stressors nonetheless crop up in the literature. The most important, not surprisingly, is exercise. Humans are built to move, so move we should. The more research emerges on physical exercise, the more we find that its benefits extend far beyond our muscles and cardio-vascular systems. Exercise expands the productive capacity of our amine-producing cells, helping to inoculate us against anxiety, stress, depression and learned helplessness. It also floods our brains with what are called growth factors, and these keep existing neurons young and new neurons growing – some scientists call these growth factors 'brain fertiliser' – so our brains are strengthened against stress and ageing. A well-designed regime of physical exercise can be a boot camp for the brain. In the future, however, the advice to exercise, administered so liberally by doctors everywhere, could be made more effective by being more explicit. What type of exercise? Anabolic or anaerobic? How often? Once again, sports science could help enormously in tailoring this advice to the person receiving it.

One type of toughening regime is especially intriguing, and that is exposure to cold weather, even to cold water. Scientists have found that rats swimming regularly in cold water develop the capacity to mount a quick and powerful arousal, relying on adrenalin more than cortisol, and to switch it off just as quickly. When subsequently exposed to stressors they are not as prone to learned helplessness. Some tentative research has suggested that much the same thing occurs in humans. People who are regularly exposed to cold weather or who swim in cool water may have undergone an effective toughening regime that has made them more emotionally stable when confronted by prolonged stress. It is surmised by some researchers that the exercise itself, coupled with acute thermal demands, provides these people with an enviable pattern of stress and recovery. Perhaps the same effects could result from the Nordic practice of a sauna followed by a cold plunge.

Recall that thermoregulation represented a revolutionary advance for mammals, profoundly altering their bodies, brains, and the network of connections between the two; and that it proved particularly so for early humans, whose superior ability to cool their bodies gave them an edge on the African savannah. Some scientists have even claimed that the nervous system supporting thermoregulation in mammals laid the foundation for later systems of emotional arousal. Dienstbier has elaborated this idea, and argues that people who have developed cold tolerance may also have increased their emotional stability.

Thermal stress is a natural part of our life, so if it is eliminated a fundamental part of our physiology may atrophy. The great physiologist Walter Cannon hinted at something like this back in the 1920s. Displaying an extraordinary prescience, he worried about the advent of central heating, air conditioning and hot running water, because these conveniences threatened to deprive us of the opportunity of exercising our systems of thermoregulation. 'It is not impossible,' Cannon warned, 'that we lose important protective advantages by failing to exercise these physiological mechanisms, which were developed through myriads of generations of our less favored ancestors. The man who daily takes a cold bath and works until he sweats may be keeping "fit", because he is not permitting a very valuable part of his bodily organization to become weakened and inefficient by disuse.' Today we may be paying a high price for our modern comforts. In fact, Cannon's fears of a decline in fitness may have been justified: recent evidence suggests that the widespread adoption of climate control in home, car and office may be one cause of the current obesity epidemic. The disappearance of thermal stress from our lives may have another unintended consequence: it may have largely eliminated a valuable toughening process.

It is too early in this research to recommend any particular toughening regime as a way for risk-takers in the financial world and elsewhere to build up a resilience to the stress that inevitably accompanies their job. However, I think financial institutions should take seriously the fact that a trader's ability to handle risk involves a lot more than a knowledge of probability, macroeconomics and formal finance. Traders need to be trained so they can recognise and handle the physiological changes resulting from their gains and losses, and from market volatility. These training regimes will have to be designed in such a way that they access the primitive brain, not just the rational cortex. Since the body profoundly influences sub-cortical regions of the brain, the new training programmes may turn out to involve a lot more physical exercises than they do at present. Banks and hedge funds could potentially learn from the programmes of top-class athletes, for they are the people with the most experience of controlling their hormones and emotions in the interests of optimising performance.

LEARNING TO LISTEN

Is there anything we can do once exhaustion, fatigue, anxiety or stress have set in? To answer this question we must bear in mind that these conditions are messages sent from our body telling us what actions we should take, and we need to understand what they are saying. Quite often, though, we misunderstand these messages. A telling example can be found in our understanding of mental fatigue. Common sense tells us that it is a state of exhaustion, in which we have quite simply run out of energy, like a car running out of gas. The recommendation that naturally follows is a rest or vacation to replenish our energy reserves. Exhaustion of this kind certainly occurs. Run a marathon, and chances are you end up in a state of exhaustion; pull an all-nighter, and chances are you need some sleep. But more often than not, this is not the cause of mental fatigue. Often, mental fatigue disappears if we merely change activities, and that would not happen if we had exhausted our fuel.

A recently developed model in neuroscience provides an alternative explanation of fatigue. According to this model, fatigue should be understood as a signal our body and brain use to inform us that the expected return from our current activity has dropped below its metabolic cost. The brain quietly searches for the optimal allocation of attentional and metabolic resources, and fatigue is one way it communicates its results. If we are engaged in some form of search and have not turned up any results, our brain, through the language of fatigue and distractibility, tells us we are wasting our time and encourages us to look elsewhere. The cure for fatigue, according to this account, is not a rest, it is a fresh task. Support for this idea comes from data showing that overtime work does not in itself lead to work-related illness such as hypertension and heart disease; these occur mainly if workers have no control over the allocation of their attention. Applying such a model could benefit workers and managements alike, for more flexibility in choosing what to work on, and when, could reduce worker fatigue, while management might be delighted to find that workers may be just as refreshed by a new assignment as by a vacation. This model of fatigue provides a good example of how understanding a bodily signal can alter the way we deal with it.

Novelty may thus prove rejuvenating when we are battling fatigue, but under other circumstances it can turn toxic – when, for example, we are trapped in a state of chronic stress. Berlyne's ∩-shaped hill displays how novelty and complexity beyond moderate levels can promote anxiety. If we return to chronic stress and look at the influence of novelty in this condition, we can find another example of how we frequently misunderstand the source of our problems.

In a novel situation we do not know what to expect, so our body mounts a preparatory stress response. That much is perfectly understandable. What is less obvious is that it does not seem to matter whether the novelty is welcome or dreaded, for either can exacerbate chronic stress. This conclusion emerged from a study by two psychiatrists who compiled a list of life-changing events, known as the Holmes and Rahe Social Readjustment Rating Scale, which they used to predict future illness and death. They found that all the obvious stressors, such as divorce, the death of a spouse or financial difficulties, predicted a heightened risk of illness and death. But also high on their list were more welcome changes, such as marriage, the birth of a child, a change of job or, incredibly, outstanding personal achievement. While these events were no doubt welcome, they added novelty to the lives of the recipients, and that could later take a toll on their health. Our complete unawareness of the damage being inflicted on us at such times is one reason hypertension and heart disease are called silent killers.

Findings such as these can change the way we handle chronic stress. When we are mired in stress, what we desperately need to do is minimise the novelty in our lives. We need familiarity. But quite often we seek out the exact opposite, responding to chronic stress at work, for example, by taking a vacation in some exotic place, thinking that the change of scenery will do us good. And under normal circumstances it does. But not when we are highly stressed, because then the novelty we encounter abroad can just add to our physiological load. Instead of travelling, we may be better off remaining on home turf, surrounding ourselves with family and friends, listening to familiar music, watching old films. Exercise, of course, can help, in fact there are few things better at preparing our physiology for stress, but when someone is this far into chronic stress its effects, suggests Stephen Porges, are mostly analgesic, possibly because exercise treats us to a shot of natural opioids. Again, what we really need is familiarity.

Besides dampening physiological arousal, familiarity can have another beneficial effect. It can convince our vagus nerve, that angel of mercy, to become maximally involved in our problems, take charge of our shattered body, and calm things down. The vagus has in its hands the power to slow our stressed heart, ease our breathing, settle our stomach. It can save our life. But in order to do so, it needs familiar surroundings, and more specifically the faces and voices of friends and family. As we have seen, the vagus ties together face and voice, and the brain stem regions controlling arousal. Familiar voices and happy faces let our brain stem know that fight-or-flight is not needed, so the vagus informs the body that it can stand down from a state of high alert. If you are blessed with a calm family and friends whose fortunes are uncorrelated with your own, it can help enormously in times of stress just to look into their faces and listen to their happy voices, rather than staring at your BlackBerry, gnawing on your fingernails and ruminating over past outrages.

The vagus remains something of a mystery, so we are not yet in a position to fully understand how else we can engage it. Vagal nerve stimulation (VNS), in which an electronic device is implanted in the chest to artificially stimulate the nerve, has met with success in treating otherwise intractable depression and chronic pain, although how exactly it achieves its results is not fully understood. But we do know a few things.

As we have seen, the dive reflex, triggered when you splash or submerge your face in cold water, can engage the vagus and slow the heart, breathing and metabolism. Breathing exercises, involving slow, deep breaths from the diaphragm, rather than short, shallow ones from the chest, can also bring the vagus online, as will similar practices, suggests Porges, such as 'playing wind instruments, singing, and even expanding the duration of phrases when talking – all will have a profound effect on vagal influences to the heart'. The calming effect of controlled breathing is a well-known biofeedback practice. It also forms an important part of yoga, meditation and some Eastern religions, especially the Buddhist art of mindfulness, in which mind and body are united through a focused attention on breathing. This exercise may have further benefits: the neuroscientist Read Montague and his colleagues have found that people practising Buddhist meditation engaged their gut feelings more than others, and as a result made more rational choices in a financial decision-making task. Research into stress, the vagus and gut feelings is in this way building a bridge between East and West.

We have only just started to understand and harness the power of the vagus nerve. Can we increase vagal tone? Can we train this nerve to come online sooner and act more powerfully, so we rely less on the metabolically expensive fight-or-flight response, or worse, the full stress response, with its attendant high levels of cortisol? Given the power and the range of action of the vagus, finding the answers to these questions represents something of a Holy Grail for stress research.

STRESS IN THE WORKPLACE

Beyond individual toughening regimes, are there any objective changes we could make in the workplace to reduce stress? Novelty, uncertainty and uncontrollability seem endemic to the markets themselves, so how could banks minimise these conditions? Should they even try? Maybe not. At moderate levels, uncertainty in markets provides a spark igniting the risk-taking that is the allotted role of a bank. But that is not what I am concerned with here. What concerns me is the debilitating effects of chronic stress on health, risk-aversion and, beyond that, financial market instability. Can management learn anything from physiology to help it alleviate these pathological conditions? I believe it can.

The Whitehall Studies mentioned in the previous chapter looked at the health consequences of job insecurity and uncontrollability in the British civil service, and found that the uncertainty created among employees led to a noticeable increase in hypertension, cholesterol levels and heart disease. My colleagues and I similarly found that market uncertainty, as measured by volatility and uncontrollability in traders' P&Ls, had a very powerful effect on their cortisol levels.

Reducing uncertainty and giving people even a modicum of control can have noticeable health effects. Doctors have found this with patients experiencing pain, who suffered even more when they did not know when they would get their pain medication. In a radical experiment, a few doctors gave patients the ability to administer the painkillers themselves, and the amount used actually dropped. Removing the uncertainty and uncontrollability had the effect of reducing the need for painkillers. Pain is a signal, telling us to keep off damaged tissue; and during stress, it is reasonable to assume, the signal is warning us that we are in greater danger of doing more damage. Remove the stress, and perhaps the signal no longer needs to be quite so strong. Patient-controlled analgesia, as it is called, has now become standard practice in many hospitals.

It may not be possible to eliminate uncertainty and uncontrollability in the workplace, so it behoves workers to reduce uncertainty and establish as much control as possible outside the workplace, in their personal lives. We may not be able to control the financial markets, but we can exert some control over our own bodies, what we eat, how often we go to the gym, who we spend time with, etc. By doing so we gain a toehold amidst the chaos, and this can convince us – delude us, if you will – into believing we are in charge. It may be an illusion, and it may not stop you from losing money or even being fired, but it can help reduce the long-term damage to your body.

Another powerful antidote to the physical damage wrought by uncertainty and uncontrollability is social support. A circle of close friends and family, and a supportive management team at work, can be a particularly potent force in mitigating the damage of stress. Just how potent became apparent from a study on stress and mortality conducted in Sweden. The researchers interviewed 752 men, asking them to indicate how many serious life events had occurred to them recently, such as divorce, being fired or financial troubles. Seven years later, the researchers followed up on the men. The death rate among those who had reported being chronically stressed was three times higher than among those who had reported no stress. However, among the men who did report stress, those who had a supportive circle of friends and family showed no correlation at all between the stressful life events and increased mortality.

Equally effective in combating the effects of uncertainty and uncontrollability in the workplace is a policy of devolving control. In their pioneering book *Healthy Work*, mentioned in the previous chapter, Theorell and Karasek investigated a highly influential management model in which specialised workers mechanically execute plans handed down by upper management. Most jobs in companies adhering to this model had a high workload and low control; they also had a high incidence of stress-related illness. Theorell and Karasek asked whether illness in the workplace is the inevitable price we pay for higher profits. They concluded that it is not, that a healthy worker is a productive one. Moreover, a healthy worker is one with far lower medical costs, and these can add up to substantial savings both for the employing company and for the economy as a whole. Their research, and other studies like it, suggests that levels of workplace stress and worker health should become goals of management just as much as short-term profits. Theorell and Karasek cite the case of a Volvo car plant in Sweden which devolved control over many details of production to small groups of workers, and found that the incidence of stress-related illness dropped noticeably.

In banks the stress of uncertainty and uncontrollability stems from the market and P&L, but also, as in other workplaces, from the chain of command. Some of these sources of uncertainty are easily minimised. Certain hedge funds, for example, recognise that trading involves two separable stages, conceiving a trade and executing it, and that responsibility for the latter can be removed from the decision-makers and given to an execution desk. Many traders miss the execution side of trading, but managers who have tried this innovation believe it reduces stress and improves decision-making. Banks employ another tactic to reduce their traders' stress and feelings of uncontrollability. When financial crises hit, traders often end up with large positions in risky assets, such as mortgages or junk bonds, which they cannot sell. The losing positions hang around their necks like a deadweight, making daily trading all but impossible. Upper management often removes these positions from the accounts of individual traders so they can concentrate on new business.

Within the banks the most potent source of novelty, uncertainty and uncontrollability is, however, managerial instability. When crises hit and traders lose money, they will inevitably suffer stress, but this stress is nothing compared to that generated by the rumours that swirl around a bank about layoffs, management shake-ups, someone else taking over your responsibilities, and so on. These, I believe, cause the bulk of stress in a bank; and unfortunately this type of stress usually occurs just when we can least handle it, during a crisis. If we want to stabilise risk preferences in the financial sector, or at least those shifts in risk preferences that stem from stress-related physiological changes, we should reduce as much as possible the stress coming from management. The market may not be controllable, but stress coming from managerial instability can be, at least much more than it is at present.

Middle management during crises often acts like those dominant monkeys that, when subjected to stress, take to bullying juniors. Senior management should therefore restrain middle managers from venting their frustration on traders (and salespeople), no matter how hard that is, no matter how deserving of termination they may be. If this sounds as if I am arguing for a cosy and supportive atmosphere for traders who may well have helped blow up our financial system, I am not. I am concerned rather with stabilising risk preferences among a financial community that may develop, as a crisis wears on, into a clinical population. Once that happens, the entire economy suffers.

The most effective way of reducing stress in the financial world, however, may be to stabilise the nature of bankers' and traders' careers. We have to make employment in the financial sector more akin to building a career than to elbowing for room at the roulette wheel. In an ideal bank we would find bonus incentives, risk-management schemes and hiring policies designed to counteract the instability in our biology, smoothing out the waves. But unfortunately what we find is precisely the opposite – bonuses, risk limits and hiring practices that operate in a powerfully pro-cyclical manner, expanding during the boom, contracting during the bust. To see how this works, we can begin by taking a closer look at bonus payments. Consider, therefore, the following scenario involving two traders, let us call them Tortoise and Hare.

Tortoise makes his employing bank $10 million a year for five years, and receives a yearly bonus of $1 million. Hare makes $100 million a year for four years, receives a yearly bonus of $20 million – the higher percentage payout is to keep this star from leaving for a hedge fund – but in the fifth year loses $500 million and receives no bonus. Despite Hare's huge loss, he does not have to repay his past bonuses. Doing our sums, we find that at the end of five years Tortoise has made the bank $50 million and has been paid $5 million, while Hare has lost the bank $100 million yet has pocketed $80 million.

Now ask yourself, which of these traders would you rather be? And do not assume Hare is out of a job. Losing a lot of money is often taken as a sign that you are a 'hitter', and can be rewarded with a job offer from another bank or hedge fund. If you are going to lose money in the finance industry, lose big.

The fable of the two traders is simple enough, yet the strategic calculation underlying Hare's choice of trading style has acted like an acid eating away at the integrity of our financial system. Anyone taking risks soon realises that their interests lie in maximising the volatility of their trading results and the frequency of their bonus payments. This strategy increases their chances of being paid at 'high-water marks', like the years when Hare made the bank $100 million.

And what works for traders also works for their managers, and even the bank's CEO: all have concluded that to maximise their long-term wealth they should focus on short-term profits. Besides, if next year all the Hares blow up, then all the banks blow up together, and no single person or bank looks bad.

Perhaps most dangerous of all, though, is that with the backing of management, traders opting for Hare's strategy are given the freedom to increase their risk just when the system least needs it, during bull markets. In this way risk-management and the insidious logic of bonus calculations work to amplify the destabilising biology of risk-taking.

One way to tame these tidal waves of risk-taking and keep life on the trading floor safely between the tidelines is to institute a bonus scheme that pays traders once a business cycle (approximately four to five years), instead of once a year. If traders are profitable over a few years, they can begin to draw on their bonus pool. But if, like Hare, they lose all their profits after a few years, then they lose all previous bonuses. Banks could also increase the amount of the deferred bonuses the longer a trader remained profitable, effectively paying, say, 5 per cent on one-year returns, 7 per cent on two-year, 10 per cent on three-year, and so on.

Risk managers for their part could spend more of their time reining in the trading stars, even pulling them off the floor for days at a time to give them a physiological cooling-off period, much like a rain break during a tennis match. The 2002 Sarbanes–Oxley Act in the US, designed to improve corporate governance and financial disclosure, encourages mandatory vacations during which there is to be no contact with the office. These vacations may have the inadvertent effect of breaking up physiological feedback loops and returning a risk-taker's body to normal.

Hiring policies too should be altered. When markets are roaring, banks hire as if growth will continue exponentially; and when the inevitable downturn comes, they fire just as indiscriminately. I have heard the hiring policies at banks referred to as a rotating door. I have also heard managers refer to trading floors, with their rapid turnover, as self-cleaning ovens. These practices work to exaggerate the stress once a crisis hits.

If, however, bankers were hired in smaller numbers and, should they work out, received a long-term commitment from their employers, and if bonuses were calculated over a longer period of time, we would probably find that trading losses were a lot smaller, volatility in earnings reduced, and the need for layoffs during bad patches largely eliminated. If traders knew their long-term interests were served by prudent rather than reckless risk-taking, and if they further knew that prudent risk-taking guaranteed them secure employment, then I am willing to bet we would find much less irrational pessimism and risk-aversion just when the economy needs risk-takers.

In short, nature and nurture, biology and management, both contribute to creating financial crises, and both need to be addressed if we are to mitigate them. I and my colleagues found evidence that management has more than enough clout to tame the beasts lurking within risk-takers. For we found, in one of the studies previously mentioned, both high testosterone levels and high Sharpe Ratios (i.e. high profits relative to the risk taken) in the same traders. How can that be? If testosterone increases a trader's appetite for risk, could it not just as easily lead to rogue trading? Probably. But on the floor where we carried out the study, the managers employed a draconian risk-management system – pulling the plug quickly on losing trades and telling traders to sit on their hands when not in the zone – combined with a profit-sharing scheme, and this had effectively harnessed the traders' risk appetite. On this trading floor, nature and nurture combined to encourage prudent and profitable risk-taking. In trading, as in sport, we concluded, biology needs the guiding hand of experience and well-structured incentives.

There is a further step banks could take to help improve the health of their workers, and through this to stabilise risk-taking. That is by participating in what are called 'wellness programmes'. A wellness programme is a form of preventive medicine, with the difference that the clinic running the programme is often located in a gym or in the workplace. One or two private medical companies in the UK and the US have combined the venues, placing medical clinics inside gyms, and locating these hybrid clinics inside offices. By doing so they have given employees the opportunity of seeing at one and the same time a personal trainer, a physiologist and a doctor. These programmes offer a unique way of surveying an employee's life, from workplace to home to recreation to eating habits, in a manner that can help co-ordinate stress reduction. They also enable the medical company to spot trends in the health of a company's employees. If it sees a high incidence of a certain musculo-skeletal disorder, it can look on the work floor for the cause. If it finds a high incidence of stress-related illness, it can similarly go looking for its cause.

In short, once we come to understand the signals our bodies send us, including fatigue and stress, there is a great deal we as individuals can do to toughen ourselves against their ravages, and we as managers can do to minimise their impact. It is wise and far-seeing managers who put health and stabilised risk preferences at the top of their company's agenda.

NINE

From Molecule to Market

MANAGING THE BIOLOGY OF THE MARKET

Financial crises are now occurring with alarming frequency, and with far greater severity than at any time since the Crash of 1929. This instability has been caused mostly by fundamental changes in the markets – historically low real interest rates; financial deregulation; low margin requirements and high leverage; the opening up of vast new markets in Asia and the emerging economies; and lastly, but importantly, the decline of partnerships on Wall Street, in the City of London and elsewhere, with an attendant shift in priorities from long-term to short-term profits. But the bull and bear markets resulting from these changes have been grossly exaggerated by the irrational exuberance and pessimism of risk-takers. And these, I have argued, are biological reactions to conditions of above-average opportunity and threat. Hormones – and the cascade of other molecular signals hormones trigger – may build up in the bodies of traders and investors during bull and bear markets to such an extent that they shift risk preferences, amplifying the cycle.

Indeed, under the influence of pathologically elevated hormones, the trading community at the peak of a bubble or in the pit of a crash may effectively become a clinical population. In this state it may become price and interest-rate insensitive, and contribute greatly to the violence and intractability of runaway markets, to what Nassim Taleb has called 'Black Swan' events. Perhaps this explains why central banks have met with such little success in arresting a bull market or placing a safety net under a crashing one. When building models of the risks facing a bank or an economy, risk managers and policy-makers should therefore bear in mind the likely clinical state of the trading community under extreme scenarios. Should the stock market drop 50 per cent in the next year, for example, then it is safe to assume the trading community will be in a traumatised state and may not respond to lower interest rates.

One economist who fully understood the challenges for policy of non-rational decision-making was Keynes. He insightfully described how 'animal spirits' drive investment and market sentiment, but he lacked any training in biology, so he never attempted to specify what exactly these animal spirits were. Nonetheless, as animal spirits bulked larger in his thinking, the less faith he had in the rate of interest as a tool for managing the economy. That is one reason he came to believe in fiscal policy, the state taking over the role of stabilising an economy that can no longer do it on its own. Keynes harboured doubts about the ideal of a life guided by, and public policy directed at, rational choice. He once humorously hinted at these doubts when recounting a conversation he had with his friend Bertrand Russell, the arch-rationalist philosopher. Russell, he recalled, claimed that the problem with politics was that it is conducted irrationally and that the solution was to start conducting it rationally. Keynes dryly commented that conversations along these lines were really quite boring. And they are, still, today.

How, then, can we deal with irrational exuberance and pessimism? Can bank and fund managers and central bankers manage the biology of risk-takers? Here we are off any map drawn by rational-choice theory. We inhabit a culture dominated by Platonic and Cartesian ideals according to which reason is the ultimate arbiter of our deci-sions and behaviour. If we are indeed built this way, then to cure irra-tional behaviour, risk managers and policy-makers need only provide people – in this case traders and investors – with more information, or help them draw correct conclusions from the information they already have. Here the proposed cure for irrational risk-taking is a talking cure. Alternatively, governments and central banks could change prices in the market, such as the rate of interest, and let rational economic agents reallocate their spending and investing dollars accordingly. Unfortunately, the policies following from rational-choice theory have not been terribly successful in stabilising the market; and the ideal of rational choice itself has prevented us from building up any skill in dealing with a human biology run rampant, either at an individual level or at the level of policy.

And this challenge is not isolated to the financial markets, for it occurs elsewhere as well. David Owen, the senior British politician and neurologist mentioned above, has been studying this problem in the political world. During his lengthy career in the House of Commons and then the House of Lords, stretching from the 1960s to the present, Owen has observed many political leaders succumb to something very like irrational exuberance, and their resulting hubris often wreaked havoc on the country. Owen recognises that this syndrome, acquired while exercising power, presents a conundrum for political theory: how do we protect the country from leaders who develop what amounts to a mental illness while holding office? Owen's concerns echo those of central bankers who similarly face the problem of managing and containing the damage done by an unbalanced biology in the markets. Again, there is not much in our canon of economic and political theory to help us deal with these problems.

Yet recent research in neuroscience and physiology has suggested that there is a great deal we can do. In the previous chapter we looked at some research which indicated ways in which individuals could recognise, control and toughen themselves against stress and hormonal imbalances. And we looked as well at how management could help dampen the stress response in the workplace. Trading managers could further recognise that the training and managing of risk-takers requires a lot more than imparting vast quantities of information to them; it crucially needs the training of skills. Risk managers too should rely as much on behavioural observation of risk-takers – one preferably informed by a basic knowledge of physiology – as on quantitative metrics. The reliance on metrics alone proved spectacularly unsuccessful in predicting and managing the credit crisis.

There may be another way of defusing the explosive mix of hormones and risk-taking in the market, and that is by changing its biology.

WOMEN IN THE FINANCIAL WORLD

How do we do that? If a bull market is amplified by a testosterone feedback loop among traders and investors – and my own experience in trading, my and my colleagues' experiments with traders, and other researchers' studies of hormones strongly suggest it is – does this mean bubbles are largely a male phenomenon? If they are, could instability in the markets be reduced by having more women and older men employed in them? We know that market stability is served by having a diversity of opinions – we want some people buying while others sell – so perhaps the same could be said of biology, that market stability needs biological diversity. And women and older men have very different biologies from young men.

Consider older men. Hormones change over the course of a man's life (over a woman's too). Testosterone levels in men rise until their mid-twenties, then go into a slow decline that accelerates after the age of fifty. At the same time, cortisol levels drift upwards. As they age, men may therefore become less and less susceptible to the testosterone feedback loops that I have argued can morph risk-taking into risky behaviour. In addition to their altered biology, older men bring to a bank or fund a lifetime of valuable experience. They have seen bad things happen – the Crash of '87, say, or the Savings & Loan crisis of the late 1980s and early 1990s, when hundreds of US banks became insolvent – so they are less likely to jump into risks before thinking through a wide range of possible outcomes. Yet trading floors are traditionally hostile to older traders, perhaps because their slower reaction times or their more cautious attitude is misinterpreted as fear. But there is little evidence that age impairs the judgement of investors, or their ability to take risk. In fact most legendary investors, such as Warren Buffett and Benjamin Graham, achieved their status at a later stage in their lives, not as young men.

Women, for their part, have very different biologies from men. They produce on average about 10–20 per cent of the amount of testosterone as men, and they have not been exposed to the same organising effects of prenatal androgens, so they too may be less prone to the winner effect than young men. Women's stress response also differs substantially from men's. One psychologist, Shelley Taylor, and her colleagues have in fact argued that the fight-or-flight reaction is more of a male response, and is not the default reaction to threat for women in quite the same way. A woman will indeed experience fight-or-flight if faced by a grizzly bear, just as a man will; but Taylor thinks that a woman's natural reaction to threat, at least within the social situations which are today our normal environment, is what she calls the 'tend-and-befriend' reaction, an urge to affiliation. She reasons that if you have children to care for, tend-and-befriend makes more sense than launching into a fist-fight or running away.

As for their long-term stress response, women on average have the same levels of cortisol as men, and these are equally volatile. But research has found that women's stress response is triggered by slightly different events. Women are not as stressed by failures in competitive situations as are men; they are more stressed by social problems, with family and relationships. What all these endocrine differences between men and women add up to is the following: when it comes to making and losing money women may be less hormonally reactive than men. Their greater numbers among risk-takers in the financial world could therefore help dampen the volatility.

A question remains: if women could have such a tonic influence on the markets, why are there so few women traders? Why are women not pushing their way onto the trading floors, and why are banks and hedge funds not waving them in? Women make up at most 5 per cent of the traders in the financial world, and even that low number includes the results of diversity pushes at many of the large banks. The most common explanations ventured for these numbers are that women do not want to work in such a macho environment, or that they are too risk-averse for the job.

There may well be a kernel of truth to these explanations, but I do not place much stock in them. To begin with, women may not like the atmosphere on a trading floor, but I am sure they like the money. There are few jobs that pay more than a trader in the financial world. Besides, women are already on the trading floor: they make up about 50 per cent of the sales force, and the sales force sits right next to the trading desks. So women are already immersed in the macho environment and are dealing with the high-jinx; they are just not trading. Also, I am not convinced women are as easily put off by a male environment as this explanation assumes. There are plenty of worlds once dominated by men that have come to employ more women: law and medicine, for example, were once considered male preserves but now have a more even balance between men and women (although admittedly not at the top echelons of management). So I am not convinced by the macho environment argument.

What about the second-mentioned explanation, that men and women differ in their appetite for risk? There have been some studies conducted in behavioural finance which suggest that on computerised monetary choice tasks women are more risk-averse than men. But here again, I am not entirely convinced, because other studies, of real investment behaviour, show that women often outperform men over the long haul, and such outperformance is, according to formal finance theory, a sign of greater risk-taking. In an important paper called 'Boys will be Boys', two economists at the University of California, Brad Barber and Terrance Odean, analysed the brokerage records of 35,000 personal investors over the period 1991–1997 and found that single women outperformed single men by 1.44 per cent. A similar result was announced in 2009 by Chicago-based Hedge Fund Research, which found that over the previous nine years hedge funds run by women had significantly outperformed those run by men.

Barber and Odean traced the women's outperformance to the fact that they traded their accounts less. Men on the other hand tended to overtrade their accounts, a behaviour the authors take as a sign of overconfidence, a conviction on the part of the men that they can beat the market. The trouble with overtrading is that every time you buy and sell a security you have to pay the bid–offer spread plus any commission, and these costs add up so quickly that they substantially diminish returns. Is the superior performance of women risk-takers due to their lower transactions costs? Or is part of it due to higher risk-taking? Or perhaps to better judgement? How can we reconcile the experimental findings that women are more risk-averse with the data on their actual returns, which suggests either greater risk-taking or better judgement? There is a clue that may help solve this mystery.

As mentioned, women make up about 5 per cent of an average trading floor. But these numbers change dramatically when we leave the banks and visit their clients, the asset-management companies. Here we find a much higher percentage of women. The absolute numbers are not large, because asset managers employ much fewer risk-takers than banks; but at some of the big asset-management companies in the UK women make up as much as 60 per cent of the risk-takers. This fact is, I believe, crucial to understanding the differences in risk-taking between men and women. Asset management is risk-taking, so it is not the case that women do not take risks; it is just a different style of risk-taking from the high-frequency variety so prevalent at the banks. In asset management one can take time to analyse a security and then hold the resulting trade for days, weeks or years. So the difference between men's and women's risk-taking may be not so much the level of risk-aversion as it is the period of time over which they prefer to make their decisions.

Perhaps men have dominated the trading floors of banks because most of the trading done on them has traditionally been of the high-frequency variety. Men love this quick decision-making, and the physical side of trading. But do trading floors today really need so many of these rapid-fire risk-takers? Banks certainly do need them; but with the advent of execution-only boxes of the sort discussed in Chapter 2, we are now in a position to disaggregate the various traits required of a risk-taker – a good call on the market, a healthy appetite for risk, and quick reactions – and let computers provide the quick execution. Increasingly, all that is required of risk-takers is their call on the market and their understanding of risk once they put on a trade; and there is no reason to believe men are better at this than women. Importantly, the financial world desperately needs more long-term, strategic thinking, and the data indicate that women excel at this. As banks, hedge funds, and asset-management companies assess their current needs and more data emerges on the performance of women risk-takers, the financial institutions will come, I believe, to hire more and more women.

Besides letting the market take its natural course, there is a policy that could hasten the hiring of women. That is to alter the period of time over which a risk-taker's performance is judged. To repeat a point made in the last chapter, the trouble in the financial world right now is that performance is measured over the short haul. Bonuses are declared yearly, and within this year there is a lot of pressure on traders to trade actively – floor managers do not like to see people sitting on their hands, even if it may be the right thing to do – and to show profits on a weekly basis. Perhaps this aggressive demand for short-term performance has prevented banks from discovering the high long-term returns of which women are capable. The solution to this problem is quite simply to judge women risk-takers – all risk-takers for that matter – over the long haul. Here again this goal could be served by calculating bonuses over the course of a full business cycle. Should we do that, we might find banks and funds not worrying too much about a slow period in a trader's returns, but only about their returns over the cycle. The market would come to value the stability and high level of women's long-term performance, and banks might naturally start to select more women traders. Affirmative action would not be needed.

There is, however, another perspective, a troubling one, from which to view this and any other proposed solution to the problem of market instability. It has been suggested to me that we should not try to calm the markets, because bubbles, while troublesome, are a small price to pay for channelling men's testosterone into non-violent activities. Andrew Sullivan, in an article mentioned in Chapter 1, has expressed a similar concern. Ruminating about the role of testosterone today, he worries that the real challenge facing us is not so much how to incorporate women more fully into society but how to stop men from seceding from it. A chilling thought. Keynes entertained somewhat the same concern, and concluded that capitalism, rather than any of the other economic systems on offer during the 1930s, was the preferred antidote to our violent urges, quipping that it is better to terrorise your chequebook than your neighbour. So maybe it is better to have testosterone vented in the markets than elsewhere.

I do not think that follows. We tend to get extreme behaviour of the sort that gives testosterone a bad name when we isolate young males. This phenomenon can be observed in the animal world, and most vividly among elephants. In the absence of elders, young male elephants go into *musth* prematurely, a condition in which their testosterone levels surge 40 to 50 times above baseline, and then they run amok, killing other animals and trampling villages. In South Africa, park rangers have found a solution to the problem: they have brought in an elder male elephant, and his presence has calmed the rogues.

This example from the animal world is admittedly more extreme than anything we find in human society, but it does dramatically illustrate the point I am making. There may be times when we want young males to be cut loose, perhaps during times of war. But when it comes to allocating the capital of society, the financial sector's allotted goal, we probably do not want volatile behaviour. We want balanced judgement and stable asset prices, and we are more likely to get these if we have the whole village present – men and women, young and old.

I like this policy, of altering the biology of the market by increasing the number of women and older men in it. It strikes me as eminently sensible, and my hunch is that it would work. To argue for it one need not claim that women and older men are better risk-takers than young men, just different. Difference in the markets means greater stability. There is, however, one point about which I have much more than a hunch, about which I am as certain as certain can be – that a financial community with a more even balance between men and women, young and old, could not possibly do any worse than the system we have now. For the one we have now created the credit crisis of 2007–09 and its ongoing aftershocks, and there is quite simply no worse outcome for a financial system.

Lastly, this policy has some decent science behind it. It provides a good example of how biology can help us understand and regulate the financial markets, and it does so, moreover, in a manner that is not in the least bit threatening.

A RETURN TO WHAT WE ONCE KNEW

It is easy to be scared by science. Novels and movies regularly portray a future in which our individuality is threatened and our dignity crushed, and more often than not it is science, rather than politics or war, that has acted as midwife to this dystopian nightmare. We find a telling example in the film *Gattaca*. A beautifully-shot and underrated film, *Gattaca* grapples with our darkest fears of genetics turned against us. Set in the not-too-distant future, it portrays a society in which a person's opportunities – their job, friends and spouse – are determined by their DNA. The main character, played by Ethan Hawke, wants the job – being an astronaut – and the girl, Uma Thurman, but does not have the right DNA, an ugly fact recorded indelibly on his identity card, as well as on every hair, fingernail and skin cell that drops off his body, ready to be collected if need be by security agents trying to verify his fitness. Similar visions of science ushering in an unwanted future can be found in tales of nuclear apocalypse, of lab-bred viruses escaping into an unsuspecting world, of computer networks rising to self-consciousness and deciding they just don't like humans.

Sometimes, though, scientific discoveries do not scare the pants off us. Sometimes they do not herald a terrifying new world, but merely reveal a tacit knowledge we have always possessed but have never been able to articulate. This tacit knowledge is shared between body and brain. Most of it, like homeostatic regulation, remains inaccessible to consciousness; some of it, like gut feelings, can be brought to the fringes of awareness; and some, like fatigue and stress, can be brought to full consciousness but are often misunderstood. We are in the strange position of creatures who on the one hand generate bodily messages intended to maintain health and happiness or prepare us for movement, but on the other sometimes do not know, or rather do not consciously know, what they mean. Where body and pre-conscious brain meet conscious brain we resemble in our split being two peoples meeting on their mutual border and each trying to communicate through a language the other only dimly comprehends. Fortunately, we are now learning to decipher these signals. Through biology we are figuring out why we are receiving these messages and how to answer them.

There is another way in which scientific discoveries can reassure us: they can remind us of a knowledge we once consciously possessed – and verbally discussed – but have forgotten. For there was a time when we candidly recognised that we are biological beings, and we thought long and hard about what this simple fact meant for ethics, politics and economics. The most notable person to do so was, as mentioned, Aristotle. In his work we find a blueprint for conceiving of life in a way that fully appreciates the role our body plays in our thinking, how it is responsible for both our agonies and our ecstasies, and how its claims must be heard at the bargaining table of policy. That is why I say that physiology and neuroscience are today returning us to an understanding of the human condition we once had, but lost, Aristotle's way of thinking being buried for millennia under layers of Platonic and Cartesian rationalism.

Aristotle created, more or less on his own, the template of the Western mind. Traces of his influence linger in every subject. Any student, no matter what department he or she studies in, will probably read something by Aristotle in their first year at university, even if it is just a paragraph or two: political science students might read the *Politics*, law students the *Ethics*, philosophers the *Metaphysics*, logicians the *Analytics* and the *Categories*, biologists the *History of Animals*, physicists and chemists a quotation or two from the *Physics*, and literature students – as well as aspiring Hollywood scriptwriters – the *Poetics*, in which Aristotle codified the structure of Western narrative. Indeed, it could be argued that within universities the departmental divisions themselves owe their very existence to the way Aristotle carved up our knowledge into its various branches, based on their subject matter and method of study.

Despite his pervasive influence, Aristotle's views on one point have been all but lost to posterity – those on mind and body. When we read in Plato that within our decaying body there burns a spark of pure reason, we are reading something so familiar that we barely take note. But when we read in Aristotle a sentence such as, 'If the eye were an animal, its soul would be seeing,' we naturally ask, what on earth does that mean? Our surprise indicates just how far we have drifted in the intervening centuries from Aristotle's way of looking at mind and body. Because for Aristotle, the two could not be separated.

On this point Aristotle's thinking is at once closer to our everyday experience of how our bodies affect our thoughts, and to cutting-edge research in neuroscience. He believed that mind is of necessity embodied, that if we did not have a body we would not, quite simply, have much to think about. Yet it is Plato's ideal of thought uncluttered by physical interference that has stood out over the centuries as the one to follow, a beacon of rationality as pure as the bleached white columns of a Greek temple. And in some ways Plato's appeal is not surprising. His ideas are clean, ordered, crystalline. In Aristotle we find none of this heavenly order. In fact, when turning from Plato to Aristotle we are often shocked at Aristotle's realism. We step from the bracing heights of Olympus into a marketplace seething with activity and sweat and emotion. Armed with a biologist's skill at observation, Aristotle set about studying all the messy details of our embodied existence, the claims made on us by desire, greed, ambition, anger, hate, as well as the pull to more noble forms of thought and behaviour, such as bravery, altruism, love and the exercise of reason. In Aristotle we find a candid and loving appreciation of the warped timber of humanity. We may thrill to the ethereal beauty of Plato's vision, but we feel at home with Aristotle.

There is another way in which the idealism–realism split between Plato and Aristotle showed itself, and that was in their political thought. Plato's *Republic* has inspired countless philosophers and political leaders over the centuries, yet his ideas of the good life tended to force people into roles for which they were ill-suited. Unearthly ideals, we have learned at great cost, too easily lead to social and political disasters. Equally, otherworldly ideals of economic rationality can too easily lead to the design of a marketplace fatally prone to financial crises.

Aristotle's realism, on the other hand, led him to recommend political institutions that accommodate actual human beings rather than idealised ones. Aristotle looked at the ways we are built, our biological details, and then judged policies and political institutions according to how well they suited our nature, how effectively they brought out the best in us and channelled what was dangerous in harmless directions. Today, through advances in neuroscience and physiology, we are rediscovering the unity of brain and body that Aristotle already understood. We should, I believe, take the next step and follow his template for conceiving a social science. With our now highly advanced understanding of human biology we are in a position to create a unified policy science, from molecule to market. If we were to do so we would find, as Aristotle did, that biology can provide us with the behavioural insights we need.

We would find something more. We would find economics beginning to merge with other subjects, such as medicine, the study of bodily and psychiatric pathologies, and with epidemiology, the study of population-wide trends in illness. The nineteenth-century German physiologist Rudolf Virchow once remarked that politics is medicine writ large, and today we could extend his dictum to economics. If the walls separating brain from body came down, so too would many barriers between subjects. With the help of human biology we could even bridge the abyss of misunderstanding that has separated what the scientist and novelist C.P. Snow once called the 'two cultures' of science and the humanities.

On a personal level, the blending of biology into our self-understanding could lead to more of Aristotle's recognition moments, and help us develop a much-needed skill at interpreting and controlling our exuberance, fatigue, anxiety and stress. Written on the Temple of Delphi was the maxim, *Know Thyself!*, and today that increasingly means know your biochemistry. Doing so turns out not to be a dehumanising experience at all. It is a liberating one.

ACKNOWLEDGEMENTS

Many friends and colleagues have helped over the years, with both research and this book. For regularly challenging my understanding of philosophy, especially German, and for providing such inspiring examples of what the plain style is capable of, I express my admiration and thanks to Michael Nedo and John Mighton. For constant support during my research, thanks go to Ed Cass, Gavin Gobby, Casimir Wierzynski, and especially to Manny Roman, who tried to solve every problem I moaned about.

Further thanks go to Gavin Poolman, John Karabelas, Vic Rao, Wayne Felson, Bill Broeksmit, Scott Drawer, Stan Lazic, Josh Holden, Sarah Barton, Kevin Doyle, Geoff Meeks, Geoff Harcourt, Jean-François Methot, Mike O'Brien, Mark Codd, Ollie Jones, Gillian Moore, Ben Hardy and Brian Pedersen.

I have gratefully received financial support for ongoing research from the Economic and Social Research Council of the UK, and the Foundation for Management Education. Mike Jones of the FME has been especially supportive.

I have asked many scientists to read sections of the book dealing with their own research: Daniel Wolpert on motor circuits; Greg Davis on visual systems; Steven Pinker on robotics; David Owen on hubris syndrome; Bud Craig on interoception; Michael Gershon on the enteric nervous system; Stephen Porges on the vagus nerve; Paul Fletcher on noradrenalin; Mark Gurnell on testosterone; Zoltan Saryai on cortisol; Richard Dienstbier on toughness; and Bruce McEwen on steroids and the brain, as well as the history of research at Rockefeller. Ashish Ranpura has read the entire manuscript. I wish to thank all these people for sharing their expertise. I have accepted all their corrections. However, many of the sections dealing with their research have been subsequently rearranged, and if in the process any mistakes have crept in, they are of course my own fault.

I would like to express my deep gratitude to a few of my close colleagues, for being such a joy to work with and for demonstrating to me the rigours of science: Linda Wilbrecht, Lionel Page and Mark Gurnell. And to Sally Coates, for her unfailing ear and magical turn of phrase.

Lastly, thanks to Georgia Garrett, Donald Winchester and Natasha Fairweather at A.P. Watt, and to my editors, Anne Collins, Louise Dennys, Nick Pearson, Robert Lacey, Eamon Dolan, Emily Graff and Scott Moyers, for assistance well beyond the call of duty. And mostly to my wife, Sarah Marangoni, on whose classical education and sure judgement I have come to rely.

JOHN COATES
Cambridge, February 2012

NOTES

INTRODUCTION

3 **Far graver crises cause less keen emotion** Winston Churchill (1930) *My Early Life.* London: Oldhams Press. p.207.

7 **every voice in the stadium, see every blade of grass** Zinedine Zidane, the great French footballer, is worth quoting: 'When you are immersed in the game, you don't really hear the crowd. You can almost decide for yourself what you want to hear. You are never alone. I can hear … someone shift around in their chair. I can hear … someone coughing. I can hear someone whisper in the ear of the person next to them. I can imagine … that I can hear the ticking of a watch.' *Zidane: A 21st Century Portrait* (2006) directed by Douglas Gordon, Philippe Parreno.

10 **swings between euphoria and fear** Greenspan, A. We Will Never Have a Perfect Model of Risk. *Financial Times* 17 March 2008.

CHAPTER 1: THE BIOLOGY OF A MARKET BUBBLE

14 **morbid fear of unemployment** Caroline Bird (1966) *The Invisible Scar.* New York: D. McKay Co.

16 **'Masters of the Universe'** Tom Wolfe (1987) *The Bonfire of the Vanities.* New York: Farrar, Straus & Giroux. The other expression that aptly caught the attitude of star bankers was 'big swinging dick', coined by Michael Lewis (1990) *Liar's Poker: Rising Through the Wreckage on Wall Street.* New York: Penguin.

17 **disorder of the possession of power … with minimal constraint on the leader** Owen, D., Davidson, J. (2009) Hubris syndrome: An acquired personality disorder? A study of US Presidents and UK Prime Ministers over the last 100 years. *Brain* 132, 1407–1410. Owen develops this theme in a fascinating book, David Owen (2008) *In Sickness and in Power: Illness in Heads of Government During the Last 100 Years.* London: Methuen.

19 Yale economist Robert Shiller Robert Shiller (2005) *Irrational Exuberance* 2nd ed. Princeton University Press.

20 catastrophic economic and political consequences Randolph M. Nesse (2000) Is the market on Prozac? *The Third Culture.*

23 renowned professor at Rockefeller For a review of the earliest work done on steroid receptors in the brain, see B.S. McEwen, P.G. Davis, B. Parsons and D.W. Pfaff (1979) The Brain as a Target for Steroid Hormone Action. *Annual Review of Neuroscience* 2, 65–112.

26 *New York Times Magazine* in April 2000 Andrew Sullivan, The He Hormone. *New York Times Magazine* 2 April 2000.

29 ghost in the machine, watching and giving orders The phrase 'ghost in the machine' was in fact coined by the Oxford philosopher Gilbert Ryle when discussing Cartesian dualism in his book *The Concept of Mind* (University of Chicago Press, 1949).

30 behavioural economics See for example Richard Thaler (1994) *Winner's Curse.* Princeton University Press. Daniel Kahneman, Paul Slovic, Amos Tversky (1982) *Judgment Under Uncertainty: Heuristics and Biases.* Cambridge: Cambridge University Press. Hersh Shefrin (1999) *Beyond Greed and Fear: Understanding Behavioral Finance and the Psychology of Investing.* Boston: Harvard Business School Press.

30 we think with our body Daniel Kahneman (2011) *Thinking, Fast and Slow.* New York: Farrar, Straus & Giroux.

31 economics and the natural sciences beginning to merge A process the Harvard biologist Edward Wilson has termed 'consilience'. Wilson (1998) *Consilience: The Unity of Knowledge.* London: Little, Brown. Wilson has been criticised for pushing what some have called a Darwinian fundamentalism, a belief that all behavioural explanation will one day be reduced to biological substrates. For a criticism of Wilson see Fodor, Jerry (1998) 'Look!', *London Review of Books* Vol. 20, No. 21. I do not have a clear view on this debate, but would reiterate my earlier comment that the physiological systems that affect risk-taking act more like lobby groups, pressuring us into certain behaviour, yet do not guarantee we will comply. We retain choice to overrule their pressures.

31 no mind–body split In the *Phaedo*, 65, Plato had claimed that thought is best conducted in the absence of bodily influences. In *De Anima*, 1.ii, Aristotle on the other hand argued: 'the soul seems unable to have anything done to it, or to do anything, without the body; this is so, for instance, with regard to feeling anger, confidence, or desire, and with sensation in general. What seems most likely to be peculiar to the soul is thought; but, if even this is a kind of imagination, or at least does not occur without imagination, then not even it can occur independently of the body.' *The Philosophy of Aristotle*, ed. Renford Bambrough, trans. A. Wardman & J. Creed. Mentor Books, 1963.

CHAPTER 2: THINKING WITH YOUR BODY

35 **Evolutionary biologists ... heat exhaustion** This theory is described by, among others, Fred H. Previc (1999) Dopamine and the Origins of Human Intelligence. *Brain and Cognition* 41, 299–350.

38 **tunic-like body** Wolpert qualifies his story by pointing out that the adult tunicate retains the rudiments of an autonomic nervous system. See Mackie, G., Burighel, P. (2005) The nervous system in adult tunicates: current research directions. *Canadian Journal of Zoology* 83, 151–183. Meinertzhagen, I., Okamura, Y. (2001) The larval ascidian nervous system: the chordate brain from its small beginnings. *Trends in Neurosciences* 24, 401–410.

38 **you do not need a brain** See for example Wolpert, D., Ghahramani, Z., Flanagan, J. (2001) Perspectives and problems in motor learning. *Trends in Cognitive Science* 5, 487–494; and Wolpert, D. (2007) Probabilistic models in human sensorimotor control. *Human Movement Science* 26, 511–524.

38 **a mind on the hoof** Andy Clark (1997) *Being There: Putting Brain, Body and World Together Again.* Cambridge, MA: MIT Press. See as well Sandra Blakeslee, Matthew Blakeslee (2007) *The Body Has a Mind of its Own: How Body Maps in Your Brain Help You Do (Almost) Anything Better.* New York: Random House. For a review of issues in embodied cognition see, Wilson, M. (2002) Six views of embodied cognition. *Psychonomic Bulletin and Review* 9, 625–636.

39 **no one has yet figured out how we do it** Stephen Pinker (1999) *How the Mind Works.* New York: Norton. pp. 4–11.

39 **dexterity of an eight-year-old child** The robot Asimo built by Honda is, however, getting close. For a review of relevant issues in neuroscience and robotics see, Chiel, H., Beer, R. (1997) The brain has a body: adaptive behaviour emerges from interactions of nervous system, body and environment. *Trends in Neurosciences* 20, 553–557.

39 **We may have a larger prefrontal cortex relative to brain size than any animal** The true measure of our superior brain relative to animals is one called the encephalisation quotient.

41 **throwing a spear, or riding a horse** See for example Rickye, S., Heffner, R., Masterton, B. (1983) The Role of the Corticospinal Tract in the Evolution of Human Digital Dexterity. *Brain Behavior Evolution* 23, 165–183.

41 **modern humans may actually have had a smaller neo-cortex than the troll-like Neanderthals** See for example Anne H. Weaver (2005) Reciprocal evolution of the cerebellum and neocortex in fossil humans. *Proceedings of the National Academy of Sciences* 102, 3576–3580. In interview she has said: 'My work provides support for the hypothesis

that the human brain continued to evolve after 30,000 years ago. It also suggests that an element of that evolution involved a reduction in the relative size of the neocortex and an absolute and relative increase in cerebellar volume. Surprisingly, it looks like the neocortex of recent humans is actually smaller in proportion to the rest of the brain than it was in either Neandertals or early modern humans.'

41 **larger cerebellum … more brainpower** On the contribution of the cerebellum to cognitive function, see for example Leiner, H., Leiner, A., Dow, R. (1993) Cognitive and language functions of the human cerebellum. *Trends in Neurosciences* 16, 444–447.

45 **idling reptile** Hulbert, A., Else, P. (1981) Comparison of the 'mammal machine' and the 'reptile machine': energy use and thyroid activity. *American Journal of Physiology* 241, R350–356.

45 **proxy for survivability** Allman, J., McLaughlin, T., Hakeem, A. (1993) Brain Structures and Life-Span in Primate Species. *Proceedings of the National Academy of Sciences* 90, 3559–3563.

47 **emotional reaction** Critchley, H.D., Mathias, C.J., Dolan, R.J. (2002) Fear-conditioning in humans: the influence of awareness and arousal on functional neuroanatomy. *Neuron* 33, 653–663. Dolan, R. (2002) Emotion, Cognition, and Behavior. *Science* 298, 1191–1194.

47 **interoception, the perception of our inner world** Craig, A.D. (2002) How do you feel? Interoception: the sense of the physiological condition of the body. *Nature Reviews Neuroscience* 3, 655–666.

48 **the sense of how we feel** Bechara, A., Naqvi, N. (2004) Listening to your heart: interoceptive awareness as a gateway to feeling. *Nature Neuroscience* 7, 102–103.

48 **awareness of the overall state of our body may be found uniquely in humans** A.D. Craig (2009) How do you feel – now? The anterior insula and human awareness. *Nature Reviews Neuroscience* 10, 59–70.

48 **help us regulate our body** For this argument see as well Watt, D. (2004) Consciousness, Emotional Self-Regulation and the Brain. *Journal of Consciousness Studies* 11, 77–82.

CHAPTER 3: THE SPEED OF THOUGHT

60 **sound of the shot to reach them** This issue is discussed in Mero, A., Komi, P.V. and Gregor, R.J. (1992) Biomechanics of Sprint Running: A Review. *Sports Medicine* 13, 376–392. See as well *Reaction Times and Sprint False Starts*, http://www.condellpark.com/kd/reactiontime.htm

61 **Screams from the crowd at the blur of the gloves** Norman Mailer (1975) *The Fight*. New York: Vintage. p.174.

62 **Ali's left jab at little more than 40 milliseconds** Schmidt, R., Lee, T. (2005) *Motor Control and Learning: A Behavioral Emphasis*. Human

Kinetics Publishers. p.149. The speed of a karate punch has been measured at 11.5 metres per second. See T.J. Walilko, D.C. Viano, C.A. Bir (2005) Biomechanics of the head for Olympic boxer punches to the face. *British Journal of Sports Medicine* 39, 710–719.

62 **his success rate approaches that of many predators in the wild** I came across this interesting statistic on the website Biological Baseball (http://www.exploratorium.edu/baseball/biobaseball.html).

62 **catches its prey on average one time out of three** See for example Mech, D., Peterson, R. 'Wolf-Prey Relations', in Mech, M., Boitani, L. (eds) (2003) *Wolves: Behavior, Ecology, and Conservation.* Chicago: Chicago University Press. It should be pointed out, however, that wolves and lions are not that skilled at lone hunting, and perhaps this is why they generally hunt in packs. The stats for a cheetah and a cougar, on the other hand, show that these predators are successful more than 50 per cent of the time.

62 **before it consciously registers in the brain** Schlag, J., Schlag-Rey, M. (2002) Through the eye, slowly; Delays and localization errors in the visual system. *Nature Reviews Neuroscience* 3, 191–200. Berry, M., Brivanlou, I., Jordan, T., Meister, M. (1999) Anticipation of moving stimuli by the retina. *Nature* 6725, 334–338.

63 **while it jumps from scene to scene** Watson, T., Krekelberg, B. (2009) The Relationship between Saccadic Suppression and Perceptual Stability. *Current Biology* 19, 1040–1043.

64 **communicated by nerves to our muscles** Sigman, M., Dehaene, S. (2005) Parsing a Cognitive Task: A Characterization of the Mind's Bottleneck. *PLoS Biology* 3(2): e37.

65 **flash-lag effect** The flash-lag effect can be watched in action on the web. Put these keywords into your search engine: 'Flash-Lag Effect. From Michael's "Visual Phenomena & Optical Illusions".' See as well MacKay, D. (1958) Perceptual stability of a stroboscopically lit visual field containing self-luminous objects. *Nature* 181, 507–508. Nijhawan, R. (1994) Motion extrapolation in catching. *Nature* 370, 256–257.

65 **left behind by the fast-forwarded blue circle** A similar phenomenon outside the lab has been reported by Tom Stafford and Matt Webb: on a stormy night, when driving on a country road, you may see the tail lights of the car in front of you, but not the car itself, which may be shrouded in darkness. Should a flash of lightning illuminate the car, you would then be treated to the optical illusion of the tail lights being located halfway up the car, because they have been advanced by your brain, but the car, previously obscured, has not. This and many other effects mentioned in this chapter are described in a wonderful and fun book by Tom Stafford and Matt Webb (2005) *Mind Hacks: Tips and Tools for Using Your Brain.* Sebastopol, CA: O'Reilly Media.

66 **responding to an auditory clue** Arrighi, R., Alais, D., and Burr, D. (2006) Perceptual synchrony of audiovisual streams for natural and artificial motion sequences. *Journal of Vision* 6, 260–268. King, A. (2005) Multisensory integration: Strategies for synchronization. *Current Biology* 15, R339–R341.

66 **A ball hit for speed** See the article by Jonathan Roberts (2005) The Basic Physics and Mathematics of Table Tennis. Posted on http://www. gregsttpages.com/gttp/

66 **horizon of simultaneity** Poppel, E. (1988) *Mindworks: Time and Conscious Experience.* Boston: Harcourt Brace Jovanovich.

68 **It is … this second sight in us, that has thrown us to the ground and saved us, without our knowing how** Erich Maria Remarque (1928) *All Quiet on the Western Front*, trans. A.W. Wheen. New York: Fawcett Columbine. p.56.

68 **roughly the capacity of an ethernet connection** Kristin Koch et al. (2006) How Much the Eye Tells the Brain. *Current Biology* 16, 1428–1434.

68 **no more than about 40 bits per second actually reaches consciousness** In R. Schmidt, G. Thews (eds) *Human Physiology* 2nd ed. trans. M. Biederman-Thorson (Berlin: Springer, 1989). Zimmermann's work and the limited bandwidth of consciousness are discussed extensively in Tor Norretranders (1998) *The User Illusion.* New York: Viking.

70 **blindsight operates without us ever being aware of it** Experiments today confirming blindsight involve inviting blind patients to reach for an object in front of them. The patients invariably reply that they cannot see a thing, but they are encouraged to try anyway. They successfully reach for the object more times than would be predicted if their efforts were mere chance. There are versions of this experiment that people with normal vision can take part in; and doing so is a bizarre experience. You are asked to indicate on a screen where you think a moving or blinking object, designed so it registers just below consciousness, is located. Even though you cannot see anything, you are invited to guess, which you do more or less randomly; and then find out that you have guessed correctly more times than you would by chance alone. And you have no idea how you did it. A similar phenomenon to blindsight has been found for hearing, a phenomenon known as 'deaf hearing', in which an animal with damage to the auditory cortex nonetheless orients to sound. On superior colliculus, see Anderson. E., Rees, G. (2001) Neural correlates of spacial orienting in the human superior colliculus; *Journal of Neurophysiology* 106, 2273–2284.

70 **we often know whether we like or dislike … well before we even know what or who it is** Joe LeDoux (1996) *The Emotional Brain. The Mysterious Underpinnings of Emotional Life.* New York: Touchstone.

70 **our startle is initiated by a symmetrical expansion of a shadow in our visual field** Caviness, J.A., Schiff, W., Gibson, J.J. (1962) Persistent fear responses in rhesus monkeys to the optical stimulus of 'looming'. *Science* 136, 982–983. Rind, F., Simmons, P. (1999) Seeing what is coming: building collision-sensitive neurones. *Trends in Neurosciences* 22, 215–220.

71 **The startle … in about 100 milliseconds** Ekman, P., Friesen, W., Simons, R. (1985) Is the startle reaction an emotion? *Journal of Personality and Social Psychology* 49, 1416–1426.

71 **activated in as little as 120 milliseconds** Schmidt, R., Lee, T. (2005) *Motor Control and Learning: A Behavioral Emphasis.* Human Kinetics Publishers.

71 **computer game Tetris** Haier, R., Siegel, B., MacLachlan, A., Soderling, E., Lottenberg, S., Buchsbaum, M. (1992) Regional Glucose Metabolic Changes After Learning a Complex Visuospatial/Motor Task: a PET Study. *Brain Research* 570, 134–143.

72 **my body moves … I trust it and the unconscious mind that moves it** Ken Dryden (2003) *The Game.* New York: Wiley. p.208.

73 **complete volitional control of behavior** Loewenstein, G. (1996) Out of Control: Visceral Influences on Behavior. *Organizational Behavior and Human Decision Processes* 65, 272–292.

73 **experiments that has tormented many a scientist and philosopher** Libet, B., Wright, E.W., Feinstein, B., Pearl, D. (1979) Subjective referral of the timing for a conscious sensory experience: A functional role for the somatosensory specific projection system in man. *Brain* 102, 193–224. Libet, B., Gleason, C.A., Wright, E.W. and Pearl, D.K. (1983) Time of conscious intention to act in relation to onset of cerebral activity (readiness-potential). The unconscious initiation of a freely voluntary act. *Brain* 106, 623–642. Libet, B. (1985) Unconscious cerebral initiative and the role of conscious will in voluntary action. *Behavioral and Brain Sciences* 8, 529–566.

74 **Scientists and philosophers have proposed many interpretations of these findings** There is a good discussion of the philosophical issues raised by Libet's experiments in Daniel C. Dennett (2004) *Freedom Evolves.* London: Penguin. See as well Fahle, M.W., Stemmler, T., Spang, K.M. (2011) How Much of the 'Unconscious' is Just Pre-Threshold? *Frontiers of Human Neuroscience* 5, 120.

74 **what we have is free won't** Ramachandran, V. *New Scientist* 5 September 1988.

75 **a meticulously engineered control mechanism** Philosophers discussing neuroscience and philosophy include: Patricia Churchland (1989) *Neurophilosophy: Toward a Unified Science of the Mind-Brain.* Boston: MIT Press. Daniel Dennett (1998) *Brainchildren: Essays on Designing Minds.* Boston: MIT Press.

79 **one of the most significant changes ever to take place in the markets**
For a vision of a future economy – not just the financial markets –
dominated by autonomous computerised economic agents, see
Kephart, J. (2002) Software agents and the route to the information
economy. *Proceedings of the National Academy of Sciences* 99 Suppl 3,
7207–7213.

CHAPTER 4: GUT FEELINGS

83 **fast and slow thinking** Daniel Kahneman (2011) *Thinking, Fast and
Slow.* New York: Farrar, Straus & Giroux.

83 **locomotion and assessment** Kruglanski, A. et al. (2000) To 'Do the
Right Thing' or to 'Just Do It': Locomotion and Assessment as Distinct
Self-Regulatory Imperatives. *Journal of Personality and Social Psychology*
79, 793–815.

83 **automatic and controlled thought** Camerer, C., Loewenstein, G., Prelec,
D. (2005) Neuroeconomics: How Neuroscience Can Inform Economics.
Journal of Economic Literature 43, 9–64. The authors further divide
automatic and controlled brain processes into cognitive and emotive
ones, giving them a four-way division of brain processes.

84 **cross that would appear at differing spots and then disappear** Lewicki,
P., Hill, T., Bizot, E. (1988) Acquisition of procedural knowledge about a
pattern of stimuli that cannot be articulated. *Cognitive Psychology* 20,
24–37.

84 **disputed the supposed reliability of intuition and gut feelings** See for
example David Myers (2002) *Intuition: Its Powers and Perils.* Yale
University Press. Stuart Sutherland (2007) *Irrationality.* London: Pinter
& Martin.

84 **get us into trouble** Richard Thaler (1994) *Winner's Curse.* Princeton
University Press. Daniel Kahneman, Paul Slovic, Amos Tversky (1982)
Judgment Under Uncertainty: Heuristics and Biases. Cambridge:
Cambridge University Press.

85 **adaptations to real-life problems** Gigerenzer, G., Hertwig, R., Pachur, T.
(eds) (2011) *Heuristics: The Foundation of Adaptive Behavior.* New York:
Oxford University Press.

85 **to overcome the shortcomings of first impressions** For example, Keith
E. Stanovich (2009) *What Intelligence Tests Miss: The Psychology of
Rational Thought.* Yale University Press.

85 **naturalistic decision-making ... decisions made out in the field by
experts** Recounted in Daniel Kahneman (2011) *Thinking, Fast and Slow.*
New York: Farrar, Straus & Giroux. Ch. 22. They discuss issues raised by
Malcolm Gladwell in *Blink: The Power of Thinking Without Thinking.*
London: Little, Brown, 2005.

85 intuition is the recognition of patterns This point was originally made by Herbert Simon (1955) A behavioral model of rational choice. *Quarterly Journal of Economics* 69, 99–118.

85 Chess grandmasters ... clues on what to do next Ferhand, G., Simon, H. (1996) Recall of Random and Distorted Chess Positions: Implications for the Theory of Expertise. *Memory and Cognition* 24, 493–503.

86 Intuition cannot be trusted in the absence of stable regularities in the environment Daniel Kahneman (2011) *Thinking, Fast and Slow*. New York: Farrar, Straus & Giroux. p.241.

86 training and hard work can indeed improve your performance Robert Shiller (2005) *Irrational Exuberance* 2nd ed. Princeton University Press. Ch. 10.

87 to build algorithms to exploit these patterns This story is told in Sebastien Mallaby (2010) *More Money Than God: Hedge Funds and the Making of the New Elite*. London: Bloomsbury.

88 the gold standard among hedge funds Coates, J.M., Page, L. (2009) A Note on Trader Sharpe Ratios. *PLoS One* 4(11): e8036.

88 Sharpe ratios ... plotted ... against the number of years I argue for this measure of performance in 'Traders need more than machismo', *Financial Times* 25 November 2009.

89 the Hamlet Problem Evans, D. (2002) The Search Hypothesis of Emotion. *British Journal of the Philosophy of Science* 53, 497–509. Evans points out that the term originated with Jerry A. Fodor (1987) Modules, Frames, Fridgeons, Sleeping Dogs, and the Music of the Spheres. In Zenon W. Pylyshyn (ed.) *The Robot's Dilemma*. Ablex.

89 we rely on emotions and gut feelings The search theory of emotions originated with Ronald de Sousa (1987) *The Rationality of Emotion*. Boston: MIT Press. See as well Jon Elster (1999) *Alchemies of the Mind: Rationality and the Emotions*. Cambridge: Cambridge University Press.

90 feeling was an integral component of the machinery of reason Antonio Damasio (1994) *Descartes' Error*. New York: Putnam & Sons. p.xii.

90 Somatic Marker Hypothesis Bechara, A., Damasio, A.R. (2005) The somatic marker hypothesis: A neural theory of economic decision. *Games and Economic Behavior* 52, 336–372.

90 without the grit of somatic markers For reviews of work on emotions and economic choice see Elster, J. (1998) Emotions and economic theory. *Journal of Economic Literature* 36, 47–74. Loewenstein, G. (2000) Emotions in economic theory and economic behavior. *American Economic Review* 90, 426–432. Grossberg, S., Gutowski, W. (1987) Neural dynamics of decision making under risk: Affective balance and cognitive emotional interactions. *Psychological Review* 94, 300–318.

92 carotid artery, which supplies blood to the brain See 'Detect the Effect of Cognitive Function on Cerebral Blood Flow' in Tom Stafford and

Matt Webb (2004) *Mind Hacks. Tips and Tools for Using Your Brain in the World*. O'Reilly Media.

92 **as the machine in your head draws more fuel** Duschek, S. et al. (2010) Interactions between systemic hemodynamics and cerebral blood flow during attentional processing. *Psychophysiology* 47, 1159–1166.

92 **23 per cent more glucose into their brains than they did when at rest** Parks, R.W. et al. (1988) Cerebral metabolic effects of a verbal fluency test: a PET scan study. *Journal of Clinical Experimental Neuropsychology* 10, 565–575.

92 **this reduces our capacity for self-control** Matthew T. Gailliot et al. (2007) Self-Control Relies on Glucose as a Limited Energy Source: Willpower is More than a Metaphor. *Journal of Personality and Social Psychology* 92, 325–336. Gailliot, M., Baumeister, R. (2007) The Physiology of Willpower: Linking Blood Glucose to Self-Control. *Personality and Social Psychology Review* 11, 303–327.

93 **climax by repeated outbreaks of expression** James, W. (1884) What is an emotion? *Mind* 9, 188–205. 'Emotion Follows upon the Bodily Expression in the Coarser Emotions', in *The Principles of Psychology*. New York: Dover (1890).

94 **the feeling of an emotion is in some ways the least important part of the experience** And to Oscar Lange, a Swede who simultaneously hit upon the same theory of emotion, the theory henceforth being called the James-Lange theory of emotions.

94 **icing on the cake** Joe LeDoux (1996) *The Emotional Brain: The Mysterious Underpinnings of Emotional Life*. New York: Touchstone.

95 **but we could not actually *feel* afraid or angry** James, W. (1884) What is an emotion? *Mind 9*, 188–205.

95 **being hot under the collar or flushed with excitement, and so on** The philosopher George Lakoff discusses body metaphors in emotional language, and more generally the embodied mind. See for example *Women, Fire, and Dangerous Things: What Categories Reveal About the Mind*. Chicago: University of Chicago Press (1987).

95 **Harvard physiologist Walter Cannon** Walter Cannon (1915) *Bodily Changes in Pain, Hunger, Fear and Rage: An Account of Recent Researches into the Function of Emotional Excitement*. New York: D. Appleton & Co.

96 **vomited the half-digested contents of his stomach** *Bodily Changes in Pain, Hunger, Fear and Rage*, p.278. Cannon was here quoting Charles Darwin.

97 **slow-motion film of patients undergoing psychotherapy** Condon, W.S., Ogston, W.D. (1966) Sound film analysis of normal and pathological behavior patterns. *Journal of Nervous and Mental Disease* 143, 338–347. Condon and Ogston also observed what they called microrhythms: 'The body of the speaker dances in time with his speech.

Further the body of the listener dances in rhythm with that of the speaker!' See also Haggard, E.A., Isaacs, K.S. (1966) Micro-momentary facial expressions as indicators of ego mechanisms in psychotherapy. In L.A. Gottschalk & A.H. Auerbach (eds) *Methods of Research in Psychotherapy*. New York: Appleton-Century-Crofts.

98 **all without any awareness it has taken place** Li, W., Zinbarg, R.E., Boehm, S.G., Paller, K.A. (2008) Neural and behavioral evidence for affective priming from unconsciously perceived emotional facial expressions and the influence of trait anxiety. *Journal of Cognitive Neuroscience* 20, 95–107.

99 **the purpose of facial expressions is not so much to express feelings, as to generate them** See for example Hess, U., Kappas, A., McHugo, G., Lanzetta, J., Kleck, R. (1992) The facilitative effect of facial expression on the self-generation of emotion. *International Journal of Psychophysiology* 12, 251–265. Some of the original works on emotions and the face were Tomkins, S. (1962) *Affect, Imagery, Consciousness: The Positive Affects*. New York: Springer. Gellhorn, E. (1964) Motion and emotion: The role of proprioception in the physiology and pathology of the emotions. *Psychological Review* 71, 457–472. Izard, C. (1971) *The Face of Emotion*. New York: Appleton-Century-Crofts.

99 **may be dampening their emotional and indeed their cognitive reactions** See for example Hennenlotter, A. et al. (2009) The link between facial feedback and neural activity within central circuitries of emotion – new insights from botulinum toxin-induced denervation of frown muscles. *Cerebral Cortex* 19, 537–542. As well as Havas, D.A. et al. (2010) Cosmetic use of botulinum toxin-a affects processing of emotional language. *Psychological Science* 21, 895–900.

99 **feel the mood portrayed on their faces** Levenson, R.W., Ekman, P., Friesen, W.V. (1990) Voluntary facial action generates emotion-specific autonomic nervous system activity. *Psychophysiology* 27, 363–384.

102 **her state of arousal to the brain** Craig, A.D. (2002) How do you feel? Interoception: the sense of the physiological condition of the body. *Nature Reviews Neuroscience* 3, 655–666.

104 **do gut feelings really come from the gut?** Mayer, E. (2011) Gut feelings: The emerging biology of gut–brain communication. *Nature Reviews Neoroscience* 12, 453–466.

104 **100 million neurons** The gut contains 500 million neurons, the small intestine 100 million. Michael Gershon, personal correspondence.

104 **the Second Brain** Gershon, M.D. (1998) *The Second Brain*. New York: HarperCollins.

104 **a tunnel that permits the exterior to run right through us** *The Second Brain*, p.84.

105 **an independent nervous system** The enteric nervous system was first

discovered by William Bayliss and Ernest Starling, two British
physiologists working towards the end of the nineteenth century. This
story is recounted in Gershon's book, *The Second Brain*.

105 **more easily aroused by emotional stimuli** Vianna, E., Weinstock, J.,
Elliott, D., Summers, R., Tranel, D. (2006) Increased feelings with
increased body signals. *Social Cognitive and Affective Neuroscience* 1,
37–48.

105 **to remember where you found it** See for example Flood, J., Smith, G.,
Morley, J. (1987) Modulation of memory processing by cholecystokinin:
dependence on the vagus nerve. *Science* 236, 832–834.

106 **our body's graded response to a challenge** Reviews of the time course
of an unfolding stress response can be found in Eriksen, H.R., Olff, M.,
Murison, R., Ursin, H. (1999) The time dimension in stress responses:
relevance for survival and health. *Psychiatry Research* 85, 39–50. Robert
Sapolsky has also looked at the time course of stress hormones in
Chapter 5 of his book *Why Zebras Don't Get Ulcers*. See as well Sapolsky,
R., Romero, M., Munck, A. (2000) How do Glucocorticoids Influence
Stress Responses? Integrating Permissive, Suppressive, Stimulatory, and
Preparative Actions. *Endocrine Reviews* 21, 55–89.

106 **pattern of nervous and hormonal activation** Ekman, P., Levenson, R.,
Friesen, W. (1983) Autonomic Nervous System Activity Distinguishes
Among Emotions. *Science* 221, 1208–1210. Levenson, R. (1992)
Autonomic Nervous System Differences Among Emotions. *Psychological
Science* 3, 23–27.

106 **facial expression and so on to each situation** Rainville, P., Bechara, A.,
Naqvi, N., Damasio, A. (2006) Basic emotions are associated with
distinct patterns of cardiorespiratory activity. *International Journal of
Psychophysiology* 61, 5–18.

108 **Iowa Gambling Task** Bechara, A., Damasio, A.R., Damasio, H.,
Anderson, S.W. (1994) Insensitivity to future consequences following
damage to human prefrontal cortex. *Cognition* 50, 7–15. Antoine
Bechara, Hanna Damasio, Daniel Tranel, Antonio Damasio (1997)
Deciding Advantageously Before Knowing the Advantageous Strategy.
Science 275, 1293–1295.

109 **the result of momentary changes in the amount of sweat lying in its
crevices** Andrew Lo and Dmitry Repin have looked at this measure
along with heart rate and respiration in traders. Lo, A., Repin, D. (2002)
The Psychophysiology of Real-Time Financial Risk Processing. *Journal
of Cognitive Neuroscience* 14, 323–339.

112 **the two sides of the brain unable to communicate with each other**
Recounted in Joe LeDoux (1996) *The Emotional Brain. The
Mysterious Underpinnings of Emotional Life*. New York: Touchstone.
pp.32–33.

112 **Strangers to Ourselves** Timothy Wilson (2002) *Strangers to Ourselves: Discovering the Adaptive Unconscious.* Boston: Harvard University Press.

112 **a coherent story, a self concept** Joe LeDoux (1996) *The Emotional Brain: The Mysterious Underpinnings of Emotional Life.* New York: Touchstone. p.33.

113 **and to uncertainty and volatility in the market** Coates, J., Herbert, J. (2008) Endogenous steroids and financial risk-taking on a London trading floor *Proceedings of the National Academy of Sciences* 104, 6167–6172.

114 **professions are coming to use coaches** An interesting account of a surgeon who started using a coach – a former professor of his – can be found in Atul Gawande, 'Personal Best. Should Everyone Have a Coach?' *New Yorker* 3 October 2011.

114 **test called heartbeat awareness** Ehlers, A., Breuer, P. (1992) Increased cardiac awareness in panic disorder. *Journal of Abnormal Psychology* 101, 371–382. Dunn, B. et al. (2010) Listening to Your Heart: How Interoception Shapes Emotion Experience and Intuitive Decision Making. *Psychological Science* 20, 1–10.

114 **Experiments with heartbeat awareness** O'Brien, W.H., Reid, G.J., Jones, K.R. (1998) Differences in heartbeat awareness among males with higher and lower levels of systolic blood pressure. *International Journal of Psychophysiology* 29, 53–63. Critchley, H., Wiens, S., Rotshtein, P., Öhman, A., Dolan, R. (2004) Neural systems supporting interoceptive awareness. *Nature Neuroscience* 7, 189–195. Werner, N.S., Jung, K., Duschek, S., Schandry, R. (2009) Enhanced cardiac perception is associated with benefits in decision-making. *Psychophysiology* 46, 1–7. Crone, E. et al (2004) Heart rate and skin conductance analysis of antecedents and consequences of decision making. *Psychophysiology* 41, 531–540.

114 **heartbeat awareness is lower in people who are overweight** Cameron, O. (2001) Interoception: the inside story – a model for psychosomatic processes. *Psychosomatic Medicine* 63, 697–710.

115 **physiological monitoring … on hormones in traders** 'Financial endocrinology. Bulls at Work. To avoid bad days, financial traders should watch their testosterone levels'. *The Economist* 17 April 2008.

CHAPTER 5: THE THRILL OF THE SEARCH

123 **cancels out the sensation it expects to result** These simulations of impending movement are called forward models. At the moment their existence is a matter of hypothesis. Miall, R., Weir, D., Wolpert, D., Stein, J. (1993) Is the cerebellum a Smith Predictor? *Journal of Motor Behavior* 25, 203–216.

123 **so the tickling has no effect** Sarah-Jayne Blakemore and Chris Frith at University College London together with Daniel Wolpert devised a tickling machine to test this hypothesis. The machine consists of a lever, which you control with one hand, which moves a soft pad that tickles your other hand. When you first use the machine you have complete control over the soft pad, its movements being therefore predictable, so it produces no tickling sensation. But as time passes the linkage between lever and pad becomes looser and looser until you have no control at all and the pad moves in a manner uncorrelated with your intentions. At that point the machine tickles. Blakemore, S., Wolpert, D., Frith, C. (2000) Why can't you tickle yourself? *NeuroReport* 11, 11–16.

124 **the moth effect** See for example Brown, I. (1991) Highway Hypnosis: Implications for Road Traffic Researchers and Practitioners. In Gale, A.G. (ed.) *Vision in Vehicles III*. North Holland: Elsevier. Charles, M., Crank, J., Falcone, D. (1990) *A Search for Evidence of the Fascination Phenomenon in Road Side Accidents*. Washington DC: AAA Foundation for Traffic Safety.

125 **notice the slightest movement** Hermans, E.J. et al. (2011) Stress-related noradrenergic activity prompts large-scale neural network reconfiguration. *Science* 334, 1151–1153. Yu, A., Dayan, P. (2205) Uncertainty, neuromodulation, and attention. *Neuron* 46, 681–691.

125 **Cocktail Party Effect ... on the other side of a crowded room** Kathleen S. Lynch, Gregory F. Ball (2008) Noradrenergic Deficits Alter Processing of Communication Signals in Female Songbirds. *Brain Behavior Evolution* 72, 207–214.

125 **The body with one bound is in full readiness** Erich Maria Remarque (1928) *All Quiet on the Western Front*, trans. A.W. Wheen. New York: Fawcett Columbine. p.54.

127 **found an elegant ∩ shape** Berlyne, D.E. (1960) *Conflict, Arousal and Curiosity*. New York: McGraw-Hill. This inverted U-shaped curve was in fact first conceived in Yerkes, R.M., Dodson, J.D. (1908) The relation of strength of stimulus to rapidity of habit-formation. *Journal of Comparative Neurology and Psychology* 18, 459–482. Berlyne was the first to link the Yerkes-Dodson law, as it is called, to information theory. For a recent synthesis of information theory and

the neuroscience of arousal, see Donald Pfaff (2005) *Brain Arousal and Information Theory: Neural and Genetic Mechanisms*. Boston: Harvard University Press.

136 **it will self-stimulate until it starves** Olds, J. (1955) Reward from brain stimulation in the rat. *Science* 122, 878.

136 **and amphetamine by 1,000 per cent** Abbott, A. (2002) Addicted. *Nature* 419, 872–874. Di Chiara, G., Imperato, A. (1988) Drugs abused by humans preferentially increase synaptic dopamine concentrations in the mesolimbic system of freely moving rats. *Proceedings of the National Academy of Sciences* 85, 5274–5278.

137 **dopamine stimulates the *wanting* of juice rather than the *liking* of it** Berridge, K.C., Robinson, T.E. (1998) What is the role of dopamine in reward: hedonic impact, reward learning, or incentive salience? *Brain Research Reviews* 28, 309–369.

137 **blue sweater worn on a date** Volkow, N. et al. (2002) Nonhedonic food motivation in humans involves dopamine in the dorsal striatum and methylphenidate amplifies this effect. *Synapse* 44,175–180. Everitt, B., Robbins, T. (2005) Neural systems of reinforcement for drug addiction: from actions to habits to compulsion. *Nature Neuroscience* 8, 1481–1489.

138 **any experience ... can deliver a shot of dopamine** Horvitz, J.C. (2000) Mesolimbocortical and nigrostriatal dopamine responses to salient non-reward events. *Neuroscience* 96, 651–656. Redgrave, P., Prescott, T., Gurney, K. (1999) Is the short-latency dopamine response too short to signal reward error? *Trends in Neurosciences* 22, 146–151. Pruessner, J., Champagne, F., Meaney, M., Dagher, A. (2004) Dopamine Release in Response to a Psychological Stress in Humans and its Relationship to Early Life Maternal Care: A Positron Emission Tomography Study Using [11C]Raclopride. *Journal of Neuroscience* 24, 2825–2831. Becerra, L., Breiter, H.C., Wise, R., Gonzalez, R.G., Borsook, D. (2001) Reward circuitry activation by noxious thermal stimuli. *Neuron* 6, 927–946.

139 **Goodies don't just fall in your lap; you have to go out and find them** Gregory Berns (2006) *Satisfaction: Sensation Seeking, Novelty, and the Science of Finding True Fulfillment*. New York: Henry Holt. p. 42. Berns speculates that the close relationship between action and reward 'stems from the dominance that classical learning theory has maintained over psychology for the last seventy years'.

139 **would not walk even a short distance to obtain it** Arias-Carrión, O., Pöppel, E. (2007) Dopamine, learning and reward-seeking behavior. *Acta Neurobiologiae Experimentalis* 67, 481–488.

140 **it makes us want to repeat these actions** Wittmann, B., Daw, N., Seymour, B., Dolan, R. (2008) Striatal Activity Underlies Novelty-Based Choice in Humans. *Neuron* 58, 967–973.

140 **crave these physical activities** Robbins, T.W., Everitt, B.J. (1982) Functional studies of the central catecholamines. *International Review of Neurobiology* 23, 303–365. Robbins, T.W., Everitt, B.J. (1996) Neurobehavioural mechanisms of reward and motivation. *Current Opinion in Neurobiology* 6, 228–236.

140 **animals prefer to work for food than to receive it passively** Denny, M. (1957) Learning Through Stimulus Satiation. *Journal of Experimental Psychology* 54, 62–64. Carder, B., Berkowitz, K. (1970) Rats' Preference for Earned in Comparison with Free Food. *Science* 167, 1273–1274. Salamone, J.D., Cousins, M.S., Bucher, S. (1994) Anhedonia or anergia – effects of haloperidol and nucleus-accumbens dopamine depletion on instrumental response selection in a T-maze cost-benefit procedure. *Behavioral Brain Research* 65, 221–229.

140 **the rapid growth of dopamine-producing cells … changed history** Fred H. Previc (2009) *The Dopaminergic Mind in Human Evolution and History.* Cambridge: Cambridge University Press.

141 **spontaneous optimism rather than on a mathematical expectation** John Maynard Keynes, *The General Theory of Employment, Interest and Money.* Ch. 12. See also George A. Akerlof and Robert J. Shiller (2009) *Animal Spirits: How Human Psychology Drives the Economy, and Why it Matters for Global Capitalism.* Princeton University Press.

142 **Rat Park** Alexander, B.K., Coambs, R.B., Hadaway, P.F. (1978) The effect of housing and gender on morphine self-administration in rats. *Psychopharmacology* 58, 175–179.

142 **will actually kick their habit** Solinas, M., Chauvet, C., Thiriet, N., El Rawas, R., Jaber, M. (2008) Reversal of cocaine addiction by environmental enrichment. *Proceedings of the National Academy of Sciences* 105, 17145–17150.

144 **You feel focused, alert, alive, motivated, anticipatory** Robert Sapolsky (2004) *Why Zebras Don't Get Ulcers* 3rd ed. New York: Henry Holt. Ch. 16.

144 **a profound feeling of satisfaction** Gregory Berns (2006) *Satisfaction: Sensation Seeking, Novelty, and the Science of Finding True Fulfillment.* New York: Henry Holt.

144 **state often described by psychologists as flow** Mihály Csíkszentmihályi (1990), *Flow: The Psychology of Optimal Experience.* New York: Harper & Row.

146 **systems are activated at the same time** Quigley, K., Berntson, G. (1990) Autonomic origins of cardiac responses to nonsignal stimuli in the rat. *Behavioral Neuroscience* 104, 751–762. Berntson, G., Cacioppo, J., Quigley, K. (1991) Autonomic Determinism: The Modes of Autonomic Control, the Doctrine of Autonomic Space, and the Laws of Autonomic Constraint. *Psychological Review* 98, 459–487.

146 **heart and lungs to full speed** Another interpretation of heart slowing is that it is merely one part of a pause, occurring throughout body and brain, before we find out what is required of us and which action should be initiated. Jennings et al. likens the pause to clutching in a car. See Jennings, R., van der Molen, M. (2002) Cardiac timing and the central regulation of action. *Psychological Research* 66, 337–349.

CHAPTER 6: THE FUEL OF EXUBERANCE

151 **dopamine in the nucleus accumbens** Schroeder, J., Packard, M. (2000) Role of dopamine receptor subtypes in the acquisition of a testosterone conditioned place preference in rats. *Neuroscience Letters* 282, 17–20. Frye, C., Rhodes, M., Rosellini, R., Svare, B. (2002) The nucleus accumbens as a site of action for rewarding properties of testosterone and its 5alpha-reduced metabolites. *Pharmacology Biochemistry Behavior* 74, 119–127.

151 **making all rewards … much more thrilling** Some of this research is described by Donald Pfaff (1999) in his book *Drive: Neurobiological and Molecular Mechanisms of Sexual Motivation*. Boston: MIT Press. See as well Fuxjager, M.J., Forbes-Lorman, R.M., Coss, D.J., Auger, C.J., Auger, A.P., Marler, C.A. (2010) Winning territorial disputes selectively enhances androgen sensitivity in neural pathways related to motivation and social aggression. *Proceedings of the National Academy of Sciences* 107, 12393–12398. Caldu, X., Dreher, J. (2007) Hormonal and genetic influences on processing reward and social information. *Annals of the New York Academy of Sciences* 1118, 43–73.

151 **steroids can be addictive** Kashkin, K., Kleber, H. (1989) Hooked on hormones? An anabolic steroid addiction hypothesis. *Journal of the American Medical Association* 262, 3166–3170.

153 **involutional melancholia** Danziger, L., Schroeder, H., Unger, A. (1944) Androgen Therapy for Involutional Melancholia. *Archives of Neurology and Psychiatry* 51, 457–461. Altschule, M., Tillotson, K. (1948) The Use of Testosterone in the Treatment of Depressions. *New England Journal of Medicine* 239, 1036–1038.

155 **in 5,000 generations men will be extinct** Bryan Sykes (2003) *Adam's Curse: A Story of Sex, Genetics, and the Extinction of Men*. Oxford University Press. See as well Steve Jones (2002) *Y: The Descent of Men*. London: Little, Brown.

155 **like sleeper cells in a spy ring** Research into these two periods in a male's life has led to a particularly elegant model of androgen action known as the organisational-activational model, according to which the sensitivity of an adult male to testosterone circulating in his blood depends on how much of the hormone he was exposed to prenatally.

(Phoenix, C., Goy, R., Gerall, A., Young, W. (1959) Organizing action of prenatally administered testosterone propionate on the tissues mediating mating behavior in the female guinea pig. *Endocrinology* 65, 369–382.) If a male was exposed to a lot of testosterone then he will have dense receptor fields or more sensitive receptors, and consequently later in life he will respond more powerfully to even small increases in the hormone circulating in his blood. If he was exposed to low levels as a foetus then later in life even large increases in testosterone may have a minimal effect.

156 **'remember' previously carried sons** Williams, T. et al. (2000) Finger-length ratios and sexual orientation. *Nature* 404, 455–456.

156 **higher levels of testosterone in later-born males** Schmaltz, G., Quinn, J.S., Schoech, S.J. (2008) Do group size and laying order influence maternal deposition of testosterone in smooth-billed ani eggs? *Hormones and Behavior* 53, 82–89. Cariello, M.O., Macedo, R.H., Schwabl, H.G. (2006) Maternal androgens in eggs of communally breeding guira cuckoos (Guira guira). *Hormones and Behavior* 49, 654–662. There is some evidence for the same mechanism operating in humans. Blanchard, R. (1997) Birth order and sibling sex ratio in homosexual versus heterosexual males and females. *Annual Review of Sex Research* 8, 27–67.

165 **deep-seated fear of stocks** Ulrike Malmendier and Stefan Nagel also make this point in 'Depression Babies: Do Macroeconomic Experiences Affect Risk-taking?' *Quarterly Journal of Economics* (2011) 126, 373–416.

165 **a peak of 44 in 1999** Robert Shiller (2005) *Irrational Exuberance* 2nd ed. Princeton University Press.

166 **a large number of species** Chase, I.D., Bartolomeo, C., Dugatkin, L.A. (1994) Aggressive interactions and inter-contest interval: how long do winners keep winning? *Animal Behavior* 48, 393–400. Rutte, C., Taborsky, M., Brinkhof, M. (2006) What sets the odds of winning and losing? *Trends in Ecology and Evolution* 21, 16–21.

166 **'resource holding potential' ... animal can draw on in an all-out fight** Hurd, P. (2006) Resource holding potential, subjective resource value, and game theoretical models of aggressiveness signaling. *Journal of Theoretical Biology* 241, 639–648.

166 **chase off a larger, well-fed animal** Neat, F., Huntingford, F., Beveridge, M. (1998) Fighting and assessment in male cichlid fish: the effects of asymmetries in gonadal state and body size. *Animal Behavior* 55, 883–891.

166 **choose fights it can win** Hsu, Y., Wolf, L. (2001) The winner and loser effect: what fighting behaviours are influenced? *Animal Behavior* 61, 777–786.

167 **deter subsequent opponents from escalating an encounter** Rutte, C., Taborsky, M., Brinkhof, M. (2006) What sets the odds of winning and losing? *Trends in Ecology and Evolution* 21, 16–21.

167 **rise in their testosterone levels** Wingfield, J.C., Hegner, R.E., Dufty, A.M., Ball, G.F. (1990) The 'challenge hypothesis': theoretical implications for patterns of testosterone secretion, mating systems, and breeding strategies. *American Nauralist* 136, 829–846. Oyegbile, T., Marler, C. (2005) Winning fights elevates testosterone levels in California mice and enhances future ability to win fights. *Hormones and Behavior* 48, 259–267.

167 **camouflage breaking** Falter, C., Arroyo, M., Davis, G. (2006) Testosterone: Activation or organization of spatial cognition? *Biological Psychology* 73, 132–140. Hines, M. et al. (2003) Spatial abilities following prenatal androgen abnormality: Targeting and mental rotations performance in individuals with congenital adrenal hyperplasia. *Psychoneuroendocrinology* 28, 1010–1026. Salminen, E., Portin, R., Koskinen, A., Helenius, H., Nurmi, M. (2004) Associations between serum testosterone fall and cognitive function in prostate cancer patients. *Clinical Cancer Research* 10, 7575–7582.

167 **hormone's tendency to increase an animal's persistence** Andrew, R., Rogers, L. (1972) Testosterone, search behavior and persistence. *Nature* 237, 343–346. Archer, J. (1977) Testosterone and persistence in mice. *Animal Behavior* 25, 479–488.

167 **and fearlessness** Boissy, A., Bouissou, M. (1994) Effects of androgen treatment on behavioral and physiological responses of heifers to fear-eliciting situations. *Hormones and Behavior* 28, 66–83.

167 **the loser with lower levels** Trainor, B.C., Bird, I.M., Marler, C.A. (2004) Opposing hormonal mechanisms of aggression revealed through short-lived testosterone manipulations and multiple winning experiences. *Hormones and Behavior* 45, 115–121. Fuxjager, M.J., Oyegbile, T.O., Marler, C.A. (2011) Independent and additive contributions of post-victory testosterone and social experience to the development of the winner effect. *Endocrinology* 152, 3422–3429.

167 **intention of depriving ... of the benefits of the winner effect** Jennings, D., Carlin, C., Gammell, M. (2009) A winner effect supports third-party intervention behavior during fallow deer, Dama dama, fights. *Animal Behavior* 77, 343–348. Dugatkin, L. (1998) Breaking up fights between others: a model of intervention behaviour. *Proceedings of the Royal Society of London* B 265, 433–437.

168 **even a depressed, state** For examples and references see James M. Dabbs (2000) *Heroes, Rogues and Lovers: Testosterone and Behavior.* New York: McGraw-Hill. pp.88–89.

168 **subject to an illusion** Gilovich, T., Vallone, R. and Tversky, A. (1985)

The hot hand in basketball: On the misperceptions of random sequences. *Cognitive Psychology* 17, 295–314.

169 **winning contributes to further wins** Page, L., Coates, J. (forthcoming) The winner effect in human behaviour: quasi-experimental evidence from tennis players.

169 **winner effect in humans?** Archer, J. (2006) Testosterone and human aggression: An evaluation of the challenge hypothesis. *Neuroscience and Biobehavioral Reviews* 30, 319–345. Mazur, A., Booth, A. (1998) Testosterone and dominance in men. *Behavioral and Brain Sciences* 21, 353–397.

169 **documented in a number of sports** Booth, A., Shelley, G., Mazur, A., Tharp, G., Kittok, R. (1989) Testosterone, and winning and losing in human competition. *Hormones and Behavior* 23, 556–571. Gladue, B., Boechler, M., McCaul, K.D. (1989) Hormonal response to competition in human males. *Aggressive Behavior* 15, 409–422.

169 **such as tennis** Booth, A., Shelley, G., Mazur, A., Tharp, G., Kittok, R. (1989) Testosterone, and winning and losing in human competition. *Hormones and Behavior* 23, 556–571.

169 **wrestling** Elias, M. (1981) Serum cortisol, testosterone, and testosterone-binding globulin responses to competitive fighting in human males. *Aggressive Behavior* 7, 215–224.

169 **ice hockey** Carré, J., Putnam, S. (2010) Watching a previous victory produces an increase in testosterone among elite hockey players. *Psychoneuroendocrinology* 35, 475–479.

169 **chess** Mazur, A., Booth, A., Dabbs, J. (1992) Testosterone and chess competition. *Social Psychology* 55, 70–77.

169 **medical exams** Mazur, A., Lamb, T.A. (1980) Testosterone, Status, and Mood in Human Males. *Hormones and Behavior* 14, 236–246.

169 **substrate to winning and losing streaks** Mazur, A. (1985) A Biosocial Model of Status in Face-to-Face Primate Groups. *Social Forces* 64, 377–402.

169 **home advantage** Neave, N., Wolfson, S. (2003) Testosterone, territoriality, and the 'home advantage'. *Physiology and Behavior* 78, 269–275. Carré, J., Muir, C., Belanger, J., Putnam, S. (2005) Pre-competition hormonal and psychological levels of elite hockey players: Relationship to the 'home advantage'. *Physiology and Behavior* 89, 392–398.

170 **achieve just the right levels of testosterone** See for example William J. Kraemer, Alan D. Rogol (eds) (2005) *The Encyclopaedia of Sports Medicine. An IOC Medical Commission Publication, The Endocrine System in Sports and Exercise, Vol. 11.* Oxford: Wiley-Blackwell. Jack H. Wilmore, David L. Costill (2004) *Physiology of Sport and Exercise* 3rd ed. Human Kinetics Publishers. Per-Olof Astrand, Kaare Rodahl, Hans A.

Dahl, Sigmund B. Stromme (2003) *Textbook of Work Physiology* 4th ed. Human Kinetics Publishers. Frank W. Dick (2007) *Sports Training Principles* 5th ed. A. & C. Black Publishers Ltd.

170 **winning the upcoming game** Carré, J., Putnam, S. (2010) Watching a previous victory produces an increase in testosterone among elite hockey players. *Psychoneuroendocrinology* 35, 475–479.

170 **manufacture my own intensity thereafter** Tim Adams (2003) *On Being John McEnroe*. New York: Crown Publishers. p.52.

170 **these too can raise testosterone levels** Carney, D., Cuddy, A., Yap, A. (2010) Power Posing: Brief Nonverbal Displays Affect Neuroendocrine Levels and Risk Tolerance. *Psychological Science* 21, 1363–1368.

170 **music was a 'divine dynamite'** Quoted in Walter Cannon (1915) *Bodily Changes in Pain, Hunger, Fear and Rage: An Account of Recent Researches into the Function of Emotional Excitement*. New York: D. Appleton & Co.

171 **in others a matter of months** Chase, I.D., Bartolomeo, C., Dugatkin, L.A. (1994) Aggressive interactions and inter-contest interval: how long do winners keep winning? *Animal Behavior* 48, 393–400.

171 **elevated testosterone for several years** van Anders, S., Watson, N. (2007) Testosterone levels in women and men who are single, in long-distance relationships, or same-city relationships. *Hormones and Behavior* 51, 286–291. Mazur, A., Michalek, J. (1998) Marriage, Divorce, and Male Testosterone. *Social Forces* 77, 315–330.

171 **than people living in rural areas** Beall, C.M., Worthman, C.M., Stallings, J., Strohl, G.M., Brittenham, G.M., Barragan, M. (1992) Salivary testosterone concentration of Aymara men native to 3600m. *Annals of Human Biology* 19, 67–78.

171 **Ache of Paraguay** Bribiescas, Richard (1975) Testosterone Levels among Ache Hunter/Gatherer Men: A Functional Interpretation of Population Variation among Adult Males. *Human Nature* 7, 163–188. James M. Dabbs (2000) *Heroes, Rogues and Lovers: Testosterone and Behavior*. New York: McGraw-Hill. p.17.

172 **World Cup final between Brazil and Italy** Bernhardt, P.C., Dabbs, J., Fielden, J., Lutter, C. (1998) Changes in testosterone levels during vicarious experiences of winning and losing among fans at sporting events. *Physiology and Behaviour* 65, 59–62.

172 **experienced vicariously by observers** Oliveira, R.F., Lopes, M., Carneiro, L.A., Canário, A.V. (2001) Watching fights raises fish hormone levels. *Nature* 409, 475. Wingfield, J.C., Marler, P. (1988) Endocrine basis of communication in reproduction and aggression. In Knobil, E., Neill, J. (eds) *The Physiology of Reproduction, Vol. 2*. New York: Raven Press. pp. 1647–1677. There is also the possibility that hormones participate in mood contagion. Neumann, R., Strack, F. (2000) 'Mood Contagion': The Automatic Transfer of Mood Between

Persons. *Journal of Personality and Social Psychology* 79, 211–223.
Totterdell, P. (2000) Catching Moods and Hitting Runs: Mood Linkage
and Subjective Performance in Professional Sport Teams. *Journal of
Applied Psychology* 85, 848–859.

172 **firm in the City of London** Coates, J., Herbert, J. (2008) Endogenous
steroids and financial risk-taking on a London trading floor. *Proceedings
of the National Academy of Sciences* 105, 6167–6172.

173 **when their morning testosterone levels were low** The analysis
presented here, as in the PNAS paper, relied on a median split on
testosterone levels. The analysis can also be carried out using panel data
and here too the results are highly significant, with correlation
coefficients in the 0.36 to 0.39 range, depending on specific analysis
used, and $p < 0.01$. I would like to thank Stan Lazic for assistance with
these statistics.

174 **ratio of the two** Manning, J., Scutt, D., Wilson, D., Lewis-Jones, D.
(1998) 2nd to 4th digit length: A predictor of sperm numbers and
concentrations of testosterone, luteinizing hormone and oestrogen.
Human Reproduction 13, 3000–3004. Manning, J., Taylor, R. (2001)
Second to fourth digit ratio and male ability in sport: Implications for
sexual selection in humans. *Evolution and Human Behavior* 22, 61–69.
John T. Manning (2002) *Digit Ratio: A Pointer to Fertility, Behavior and
Health.* Rutgers University Press.

174 **2D:4D ratios predicted how long these traders had survived in the
business** Coates, J.M., Gurnell, M., Rustichini, A. (2009) Second-to-
fourth digit ratio predicts success among high-frequency financial
traders. *Proceedings of the National Academy of Sciences* 106, 623–628.

174 **exactly what it sounds like** Cohen-Bendahana, C., van de Beeka, C.,
Berenbaum, S. (2005) Prenatal sex hormone effects on child and adult
sex-typed behavior: Methods and findings. *Neuroscience and
Biobehavioral Reviews* 29, 353–384.

175 **called *hox-a* and *hox-d*** Manning, J., Scutt, D., Wilson, D., Lewis-Jones,
D. (1998) 2nd to 4th digit length: A predictor of sperm numbers and
concentrations of testosterone, luteinizing hormone and oestrogen.
Human Reproduction 13, 3000–3004. Paul, S., Kato, B., Cherkas, L.,
Andrew, T., Spector, T. (2006) Heritability of the second to fourth digit
ratio (2d: 4d): A twin study. *Twin Research and Human Genetics* 9,
215–219. Mortlock, D., Innis, J. (1997) Mutation of HOXA13 in hand-
foot-genital syndrome. *Nature Genetics* 15, 179–180.

175 **fingers, toes and penises** Kondo, T., Zakany, J., Innis, W., Duboule, D.
(1997) Of fingers, toes, and penises. *Nature* 390, 29.

175 **increasing their appetite for risk or their confidence** Booth, A.,
Johnson, D., Granger, D. (1999) Testosterone and men's health. *Journal
of Behavioral Medicine* 22, 1–19. Apicella, C., Dreber, A., Campbell, B.,

Graye, P., Hoffman, M., Little, A. (2008) Testosterone and financial risk preferences. *Evolution and Human Behavior* 29, 384–390. Rêavis, R., Overman, W. (2001) Adult sex differences on a decision-making task previously shown to depend on the orbital prefrontal cortex. *Behavioral Neuroscience* 115, 196–206. Van Honk, J. et al. (2004) Testosterone shifts the balance between sensitivity for punishment and reward in healthy young women. *Psychoneuroendocrinology* 29, 937–943. Schipper, B. (2011) Sex Hormones and Choice under Risk (forthcoming).

175 **reduce distractions from irrelevant information** Andrew, R. (1991) in *The Development and Integration of Behaviour. Essays in Honour of Robert Hinde*, ed. Bateson, P. Cambridge: Cambridge University Press. pp. 171–190.

175 **maintain search persistence** Andrew, R., Rogers, L. (1972) Testosterone, search behaviour and persistence. *Nature* 237, 343–346.

175 **speed of reactions** Salminen, E., Portin, R., Koskinen, A., Helenius, H., Nurmi, M. (2004) Associations between serum testosterone fall and cognitive function in prostate cancer patients. *Clinical Cancer Research* 10, 7575–7582.

175 **amount of risk they took in making it** Coates, J.M., Page, L. (2009) A note on trader Sharpe ratios. *PloS One* 4: e8036.

178 **venture into the open more** Beletsky, L., Gori, D., Freeman, S., Wingfield, J. (1995) Testosterone and polygyny in birds. *Current Ornithology* 12, 141. Marler, C.A., Moore, M.C. (1988) Evolutionary costs of aggression revealed by testosterone manipulations in free-living male lizards. *Behavior Ecology Sociobiology* 23, 21–26.

178 **reduced survival** Wingfield, J.C., Lynn, S., Soma, K. (2001) Avoiding the 'costs' of testosterone: ecological bases of hormone behavior interactions. *Brain Behavior Evolution* 57, 239–251. Dufty, A.M. (1989) Testosterone and survival: a cost of aggressiveness? *Hormones and Behavior* 23, 185–193.

178 **diminished need for sleep** Pope, H., Katz, D. (1988) Affective and psychotic symptoms associated with anabolic steroid use. *American Journal of Psychiatry* 145, 487–490. Pope, H., Kouri, E., Hudson, J. (2000) Effects of supraphysiologic doses of testosterone on mood and aggression in normal men: a randomized controlled trial. *Archives of General Psychiatry* 57, 133–140.

178 **can live up to 30 per cent longer** Hamilton, J.B. (1948) The role of testicular secretions as indicated by the effects of castration in man and by studies of pathological conditions and the short lifespan associated with maleness. *Recent Progress in Hormone Research* 3, 257. Hamilton, J.B. (1965) Relationship of castration, spaying and sex to survival and duration of life in domestic cats. *Journal of Gerontology* 20, 96. D. Drori, Y. Folman (1976) Environmental effects on longevity in the male rat:

Exercise, mating, castration and restricted feeding. *Experimental Gerontology* 11, 25–32.

179 **start to stack the odds against them** For a review of our research see Coates, J., Gurnell, M., Sarnyai, Z. (2010) From molecule to market: steroid hormones and financial risk-taking. *Philosophical Transactions of Royal Society* B 365, 331–343.

CHAPTER 7: STRESS RESPONSE ON WALL STREET

189 **causing bank runs and stock crashes** Charles P. Kindleberger (2000) *Manias, Panics, and Crashes: A History of Financial Crises* 4th ed. London: Wiley.

189 **and then fall until the spring** Svartberg, J., Jorde, R., Sundsfjord, J., Bønaa, K.H., Barrett-Connor, E. (2003) Seasonal variation of testosterone and waist to hip ratio in men: the Tromsø study. *Journal of Clinical Endocrinology and Metabolism* 88, 3099–3104. Stanton, S.J., Mullette-Gillman, O.A., Huettel, S.A. (2011) Seasonal variation of salivary testosterone in men, normally cycling women, and women using hormonal contraceptives. *Physiology and Behavior* 104, 804–808. Smolensky, R., Hallek, M., Smith, M., Steinberger, K. (1988) Annual variation in semen characteristics and plasma hormone levels in men undergoing vasectomy. *Fertility and Sterility* 49, 309–315.

189 **'irritable male syndrome' ... moody, withdrawn and depressed** Lincoln, G. (2001) The irritable male syndrome. *Reproduction, Fertility and Development* 13, 567–576.

189 **tendency to outperform on sunny days** Saunders, E. (1993) Stock Prices and Wall Street Weather. *American Economic Review* 83, 1337–1345. Hirshleifer, D., Shumway, T. (2003) Good Day Sunshine: Stock Returns and the Weather. *Journal of Finance* 58, 1009–1032.

190 **Seasonal Affective Disorder** Kamstra, M.J., Kramer, L.A., Levi, M.D. (2003) Winter Blues: A SAD Stock Market Cycle. *American Economic Review* 93, 324–343.

190 **testosterone levels, for these increase with sunshine** Bernstein, I.S., Rose, R.M., Gordon, T.P. (1974) Behavioral and Environmental Events Influencing Primate Testosterone Levels. *Journal of Human Evolution* 3, 517–525. Wehr, E., Pilz, S., Boehm, B., März, W., Obermayer-Pietsch, B. (2010) Association of Vitamin D Status with Serum Androgen Levels in Men. *Clinical Endocrinology* 73, 243–248.

193 **the fast, low road** Joe LeDoux (1996) *The Emotional Brain: The Mysterious Underpinnings of Emotional Life.* New York: Touchstone. Ch. 6.

201 **slowing down both heart rate and breathing** Porges, S.W., Doussard-Roosevelt, J.A., Portales, A.L., Greenspan, S.I. (1996) Infant regulation

of the vagal 'brake' predicts child behavior problems: A psychobiological model of social behavior. *Developmental Psychobiology* 29, 697–712.

202 **low blood sugar, and so on** Selye, H. (1936) A syndrome produced by diverse nocuous agents. *Nature* 138, 32. Selye more or less discovered the stress response. He tells his story in Selye, H. (1976) *The Stress of Life*. New York: McGraw-Hill.

202 **expectation of harm than to harm itself** Mason, J. (1975) A historical view of the stress field. Part I. *Journal of Human Stress* 1, 6–12. Mason, J. (1975) A historical view of the stress field. Part II. *Journal of Human Stress* 1, 22–36. See as well Arthur, A. (1987) Stress as a state of anticipatory vigilance. *Perceptual and Motor Skills* 64, 75–85.

202 **nothing in the environment presented an overt threat** Hennessey, J., Levine, S. (1979) Stress, arousal, and the pituitary-adrenal system: a psychoendocrine hypothesis. *Progress in Psychobiology. Physiological Psychology* 8, 133–178. V. Lemaire, C. Aurousseau, M. Le Moal, D.N. Abrous (1999) Behavioural trait of reactivity to novelty is related to hippocampal neurogenesis. *European Journal of Neuroscience* 11, 4006–4014.

202 **just to be ready** Erikson, K., Drevets, W., Schulkin, J. (2003) Glucocorticoid regulation of diverse cognitive functions in normal and pathological emotional states. *Neuroscience and Biobehavioral Reviews* 27, 233–246. This is a brilliant review.

202 **Uncertainty ... provoked more stress than the shock itself** Hennessey, J., Levine, S. (1979) Stress, arousal, and the pituitary-adrenal system: a psychoendocrine hypothesis. *Progress in Psychobiolology. Physiological Psychology* 8, 133–178. Levine, S., Coe, C., Wiener, S. (1989) Psychoneuroendocrinology of stress – a psychobiological perspective. In *Psychoendocrinology*, eds Brush, F.R., Levine, S. New York: Academic Press. pp.341–377.

203 **It was in the suburbs ... higher incidence of gastric ulcers** Stewart, D.N., Winser, D. (1942) Incidence of Perforated Peptic Ulcer: Effect of Heavy Air Raids, *Lancet* 1, 259.

203 **Uncontrollability ... influence on stress levels** Breier, A., Albus, M., Pickar, D., Zahn, T.P., Wolkowitz, O.M., Paul, S.M. (1987) Controllable and uncontrollable stress in humans: alterations in mood and neuroendocrine and psychophysiological function. *American Journal of Psychiatry* 144, 1419–1425. Swenson, R., Vogel, W. (1983) Plasma catecholamine and corticosterone as well as brain catecholamine changes during coping in rats exposed to stressful footshock. *Pharmacology Biochemistry Behavior* 18, 689–693.

203 **than the one with access to the bar lever** Weiss, J. (1971) Effects of coping behavior with and without a feedback signal on stress pathology

in rats. *Journal of Comparative Physiology and Psychology* 77, 1–30. This is a three-part article.

204 **physiological foundation of the derivatives market** All the traders in our study had as one of their main trading instruments the German bond future. The implied volatility of German bond futures options showed a very high correlation of $r^2=0.86$ (p=0.001) against the traders' cortisol levels. Coates, J., Herbert, J. (2008) Endogenous steroids and financial risk-taking on a London trading floor. *Proceedings of the National Academy of Sciences* 105, 6167–6172.

205 **Our scanning becomes hurried and indiscriminate, almost panicky** Aston-Jones, G., Rajkowski, J., Cohen, J. (1999) Role of Locus Coeruleus in Attention and Behavioral Flexibility. *Biological Psychiatry* 46, 1309–1320. Berridge, C.W., Waterhouse, B.D. (2003) The locus coeruleus-noradrenergic system: modulation of behavioral state and state dependent cognitive processes. *Brain Research Reviews* 42, 33–84. Keinan, G. (1987) Decision-making under stress: Scanning of alternatives under controllable and uncontrollable threats. *Journal of Personality and Social Psychology* 52, 639–644.

208 **neighbouring brain region called the hippocampus** Lupien, S.J., Maheu, F., Tu, M., Fiocco, A., Schramek, T.E. (2007) The effects of stress and stress hormones on human cognition: Implications for the field of brain and cognition. *Brain and Cognition* 65, 209–237. Roozendaal, B. (2002) Stress and memory: opposing effects of glucocorticoids on memory consolidation and memory retrieval. *Neurobiology Learning Memory* 78, 578–595.

209 **flashbulb memories** Brown, R., Kulik, J. (1977) Flashbulb memories. *Cognition* 5, 73–99. McGaugh, J.L. (2004) The amygdala modulates the consolidation of memories of emotionally arousing experiences. *Annual Review of Neuroscience* 27, 1–28. There is some evidence that flashbulb memories may not be the photographic records we imagine, but may record how we felt during a shocking experience rather than the details. For a study of flashbulb memories of 9/11 see Tali Sharot, Elizabeth A. Martorella, Mauricio R. Delgado, Elizabeth A. Phelps (2007) How personal experience modulates the neural circuitry of memories of 11 September. *Proceedings of the National Academy of Sciences* 104, 389–394.

210 **beta-blockers … may lower the risk of later panic attacks and post-traumatic stress disorder** Cahill, L., Prins, B., Weber, M., McGaugh, J.L. (1994) Beta-adrenergic activation and memory for emotional events. *Nature* 371, 702–704.

210 **recall the events that were stored under its influence** Erickson, K., Drevets, W., Schulkin, J. (2003) Glucocorticoid regulation of diverse cognitive functions in normal and pathological emotional states. *Neuroscience and Biobehavioral Reviews* 27, 233–246.

NOTES

NOTES

210 **the amygdala and the hippocampus ... are especially affected** de Kloet, E.R., Vreugdenhil, E., Oitzl, M.S., Joels, M. (1998) Brain corticosteroid receptor balance in health and disease. *Endocrine Reviews* 19, 269–301.

210 **can kill neurons in the hippocampus** Woolley, C.S., Gould, E., McEwen, B.S. (1990) Exposure to excess glucocorticoids alters dendritic morphology of adult hippocampal pyramidal neurons. *Brain Research* 531, 225–231. Starkman, M.N., Gebarski, S.S., Berent, S., Schteingart, D.E. (1992) Hippocampal formation volume, memory dysfunction, and cortisol levels in patients with Cushing's syndrome. *Biological Psychiatry* 32, 756–765.

210 **reducing its volume by up to 15 per cent** Sapolsky, R.M. (2000) Glucocorticoids and hippocampal atrophy in neuropsychiatric disorders. *Archives of General Psychiatry* 57, 925–935.

210 **blunt the impact of stress on our brains** Bruce McEwen (2002) *The End of Stress as We Know It*. Washington: Joseph Henry Press. pp.119–124.

210 **arborisation (growth of branches)** Sapolsky, R.M. (2003) Stress and Plasticity in the Limbic System. *Neurochemical Research* 28, 1735–1742.

211 **stored reactions, largely emotional and impulsive ones** Discussed in Joe LeDoux (1996) *The Emotional Brain*. New York: Touchstone. Also Corodimas, K.P., LeDoux, J.E., Gold, P.W., Schulkin, J. (1994) Corticosterone potentiation of conditioned fear in rats. *Annals of the New York Academy of Sciences* 746, 392–339. Liston, C., McEwen, B., Casey, B. (2009) Psychosocial stress reversibly disrupts prefrontal processing and attentional control. *Proceedings of the National Academy of Sciences* 106, 912–917. Arnsten, A.F. (2009) Stress signalling pathways that impair prefrontal cortex structure and function. *Nature Reviews Neuroscience* 10, 410–422. Ohira, H. et al. (2011) Chronic stress modulates neural and cardiovascular responses during reversal learning. *Neuroscience* 193, 193–200.

211 **who did find patterns in the noise** Whitson, J., Galinsky, A. (2008) Lacking Control Increases Illusory Pattern Perception. *Science* 322, 115–117.

211 **CRH in the brain instils anxiety** Korte, S. (2001) Corticosteroids in relation to fear, anxiety and psychopathology. *Neuroscience and Biobehavioral Reviews* 25, 117–142.

212 **'anticipatory angst' ... leading to timid behaviour** Schulkin, J., McEwen, B.S., Gold, P.W. (1994) Allostasis, amygdala, and anticipatory angst. *Neuroscience and Biobehavioral Reviews* 18, 385–396.

212 **danger everywhere, even where it does not exist** McEwen, B. (1998) Stress, adaptation, and disease: allostasis and allostatic load. *Annals of the New York Academy of Sciences* 840, 33–44. Schulkin, J., McEwen, B., Gold, P.W. (1994) Allostasis, amygdala, and anticipatory angst. *Neuroscience and Biobehavioral Reviews* 18, 385–396. Lupien, S.J.,

289

Maheu, F., Tu, M., Fiocco, A., Schramek, T.E. (2007) The effects of stress and stress hormones on human cognition: Implications for the field of brain and cognition. *Brain and Cognition* 65, 209–237.

213 **irrationally risk-averse** For a review of literature on anxiety, stress and risk-aversion, see Kamstra, M.J., Kramer, L.A., Levi, M.D. (2003) Winter Blues: A SAD Stock Market Cycle. *American Economic Review* 93, 324–343.

213 **learned helplessness** Kademian, S., Bignante, A., Lardone, P., McEwen, B., Volosin, M. (2005) Biphasic effects of adrenal steroids on learned helplessness behavior induced by inescapable shock. *Neuropsychopharmacology* 30, 58–66.

213 **if the door was left open** Seligman, M., Maier, S. (1967) Failure to escape traumatic shock. *Journal of Experimental Psychology* 74, 1–9. Maier, S., Seligman, M. (1976) Learned helplessness: theory and evidence. *Journal of Experimental Psychology* 105, 3–46.

214 **impaired immune system and increased illness** Segerstrom, S. (2005) Optimism and immunity: do positive thoughts always lead to positive effects? *Brain Behavior Immunity* 19, 195–200.

214 **people become more susceptible to gastric ulcers** Chronic stress and its medical consequences are brilliantly and exhaustively reviewed in Robert Sapolsky (2004) *Why Zebras Don't Get Ulcers* 3rd ed. New York: Henry Holt.

214 **other recurrent viruses, like herpes** Segerstrom, S., Miller, G. (2004) Psychological stress and the human immune system: a meta-analytic study of 30 years of inquiry. *Psychological Bulletin* 130, 601–630.

215 **more susceptible to drug addiction** Piazza, P.V., Le Moal, M. (1998) The role of stress in drug self-administration. *Trends in Pharmacological Science* 19, 67–74. Sarnyai, Z., Shaham, Y., Heinrichs, S.C. (2001) The role of corticotropin-releasing factor in drug addiction. *Pharmacology Review* 53, 209–243.

216 **signs of chronically elevated stress hormones** Robert A. Karasek, Tores Theorell (1992) *Healthy Work: Stress, Productivity and the Reconstruction of Working Life*. New York: Basic Books. See as well Kivimäki, M. et al. (2002) Work stress and risk of cardiovascular mortality: prospective cohort study of industrial employees. *British Medical Journal* 2, 857–860. Vaananen, A. et al. (2008) Lack of predictability at work and risk of acute myocardial infarction: an 18-year prospective study of industrial employees. *American Journal of Public Health* 98, 2264–2271. Kawakami, N., Haratani, T. (1999) Epidemiology of job stress and health in Japan: review of current evidence and future direction. *Industrial Health* 37, 174–186.

216 **increased incidence of stroke** Marmot, M.G., Rose, G., Shipley, M., Hamilton, P.J. (1978) Employment grade and coronary heart disease in

British civil servants. *Journal of Epidemiology and Community Health* 32, 244–249. Ferrie, J.E., Shipley, M.J., Marmot, M.G., Stansfeld, S., Smith, G.D. (1995) Health effects of anticipation of job change and non-employment: Longitudinal data from the Whitehall II study. *British Medical Journal* 311, 1264–1269. Kuper, H., Marmot, M. (2003) Job strain, job demands, decision latitude, and risk of coronary heart disease within the Whitehall II study. *Journal of Epidemiology and Community Health* 57, 147–153. Chandola, T. et al. (2008) Work stress and coronary heart disease: what are the mechanisms? *European Heart Journal* 29, 640–648.

216 **wreak on the health of workers** See for instance Cohen, S., Schwartz, J.E., Epel, E., Kirschbaum, C., Sidney, S., Seeman, T. (2006) Socioeconomic status, race, and diurnal cortisol decline in the Coronary Artery Risk Development in Young Adults (CARDIA) study. *Psychosomatic Medicine* 68, 41–50. Ariel, K., Ziol-Guest, K., Hawkley, L., Cacioppo, J. (2010) Job Insecurity and Change Over Time in Health Among Older Men and Women. *Journal of Gerontology* B 65B, 81–90. Steptoe, A. et al. (2003) Influence of Socioeconomic Status and Job Control on Plasma Fibrinogen Responses to Acute Mental Stress. *Psychosomatic Medicine* 65, 137–144.

216 **suicide all increasing during recessions** Gerdthama, U., Johannesson, M. (2005) Business cycles and mortality: results from Swedish microdata. *Social Science and Medicine* 60, 205–218.

216 **risen 47 per cent from a year earlier** Thomas Penny, Bankers Use Secret Clinics, Nurses to Beat Breakdowns. *Bloomberg* 11 July 2008.

217 **advent of the credit crisis and the subsequent recession** Economic downturn poses threat to mental health: WHO. *CBC* 10 October 2008. See as well Elizabeth Bernstein, Angst is Rising, but Many Must Forgo Therapy. *Wall Street Journal* 7 October 2008.

217 **a spike in the rate of heart attacks in London** Smolina, K., Wright, F.L., Rayner, M., Goldacre, M.J. (2012) Determinants of the decline in mortality from acute myocardial infarction in England between 2002 and 2010: linked national database study. *British Medical Journal* 344:d8059.

217 **fear and a tendency to see danger everywhere** McEwen, B. (1998) Stress, adaptation, and disease: allostasis and allostatic load. *Annals of the New York Academy of Sciences* 840, 33–44.

CHAPTER 8: TOUGHNESS

221 **stress response that can be fatally dysfunctional in modern society** Robert Sapolsky (2004) *Why Zebras Don't Get Ulcers* 3rd ed. New York: Henry Holt. See as well Chrousos, G. (2009) Stress and disorders of the stress system. *Nature Reviews Endocrinology* 5, 374–381.

222 **proportionally less conscious influence over them** Joe LeDoux (1996) *The Emotional Brain. The Mysterious Underpinnings of Emotional Life.* New York: Touchstone.

222 **endless cycle of repetition and failure** This point is nicely illustrated in Thomas Amini, Fari Lannon, Richard Lewis (2001) *A General Theory of Love.* New York: Vintage.

222 **corticosterone levels remained stubbornly elevated** Hennessey, J., Levine, S. (1979) Stress, arousal, and the pituitary-adrenal system: a psychoendocrine hypothesis. *Progress in Psychobiology. Physiological Psychology* 8, 133–178.

223 **sees in it nothing but potential harm** Blascovich, J., Tomaka, J. (1996) The biopsychosocial model of arousal regulation. In M.P. Zanna (ed.) *Advances in experimental social psychology* 29. New York: Academic Press. pp.1–51. Blascovich, J., Mendes, W.B. (2000) Challenge and threat appraisals: The role of affective cues. In J. Forgas (ed.) *Feeling and Thinking: The Role of Affect in Social Cognition.* Cambridge: Cambridge University Press.

224 **distinctive physiological state** Blascovich, J., Seery, M., Mugridge, C., Weisbuch, M., Norris, K. (2004) Predicting athletic performance from cardiovascular indicators of challenge and threat. *Journal of Experimental Social Psychology* 40, 683–688. Although see Kirby, L.D., Wright, R.A. (2003) Cardiovascular correlates of challenge and threat appraisals: a critical examination of the Biopsychosocial Analysis. *Personality and Social Psychology Bulletin* 7, 216–233.

224 **(perhaps an environment like a financial crisis)?** Charney, D. (2004) Psychobiological mechanisms of resilience and vulnerability: implications for successful adaptation to extreme stress. *American Journal of Psychiatry* 161, 195–216.

224 **hormone feedback loops between brain and body** McEwen, B.S., Weiss, J.M., Schwartz, L.S. (1968) Selective retention of corticosterone by limbic structures in rat brain. *Nature* 220, 911–912.

225 **depleted noradrenalin levels in their brains** Weiss, J., Glazer, H., Pohorecky, L., Brick, J., Miller, N. (1975) Effects of Chronic Exposure to Stressors on Avoidance-Escape Behavior and on Brain Norepinephrine. *Psychosomatic Medicine* 37, 522–534. Weiss, J., Glazer, H. (1975) Effects of acute exposure to stressors on subsequent avoidance-escape behavior. *Psychosomatic Medicine* 37, 499–521.

225 **increased immunity to the damaging effects of further stressors**
Miller, N.E. (1980) A perspective on the effects of stress and coping on
disease and health. In S. Levine, H. Ursin *Coping and Health*. New York:
Plenum Press. pp.323–353. See as well Levine, S., Coe, C., Wiener, S.
(1989) Psychoneuroendocrinology of stress: a psychobiological
perspective. In F. Bush, S. Levine (eds) *Psychoendocrinology*. New York:
Academic Press. pp.341–377.

225 **neuromodulators, and nervous-system activation** Dienstbier, R.A.
(1989) Arousal and physiological toughness: implications for mental
and physical health. *Psychological Review* 96, 84–100.

225 **and the vagus nerve** It should be said that the number of hormones and
chemicals participating in our behaviour is far larger than the few we
have discussed in this book. To give you an idea of just how many, see G.
D. Lewis et al. (2010) Metabolic Signatures of Exercise in Human
Plasma. *Science Translational Medicine* 2, 33ra37.

226 **thriving** Epel, E.S., McEwen, B.S., Ickovics, J.R. (1998) Embodying
Psychological Thriving: Physical Thriving in Response to Stress. *Journal
of Social Issues* 54, 301–322.

227 **performance has gone stale** See for example Raglin, J., Barzdukas, A.
(1999) Overtraining in athletes: the challenge of prevention. *Health
Fitness Journal* 3, 27–31. Hakkinen, K.A., Pskarinen, A., Alen, M.,
Kauhanen, H., Komi, P.V. (1987) Relationships between training
volume, physical performance capacity, and serum hormone
concentrations during prolonged training in elite weightlifters.
International Journal of Sports Medicine 8, 61–65.

227 **state of preparedness for competition** Urhausen, A., Gabriel, H.,
Kindermann, W. (1995) Blood hormones as markers of training stress
and overtraining. *Sports Medicine* 4, 251–276. Bosquet, L., Montpetit, J.,
Arvisais, D., Mujika, I. (2007) Effects of tapering on performance: a
meta-analysis. *Medicine and Science in Sports and Exercise* 39, 1358–1365.

227 **strain we experience when stressed** Seeman, T., Singer, B., Rowe, J.,
Horwitz, R., McEwen, B. (1997) Price of adaptation – allostatic load and
its health consequences. MacArthur studies of successful aging. *Archives
of Internal Medicine* 157, 2259–2268.

227 **more potent cortisol response** Dienstbier, R.A. (1989) Arousal and
physiological toughness: implications for mental and physical health.
Psychological Review 96, 84–100.

228 **depleted of its precious noradrenalin** Berridge, C.W., Waterhouse, B.D.
(2003) The locus coeruleus-noradrenergic system: modulation of
behavioral state and state dependent cognitive processes. *Brain Research
Reviews* 42, 33–84.

230 **efficient tool for conserving energy** Porges, S.W. (1997) Emotion: an
evolutionary by-product of the neural regulation of the autonomic

nervous system. In C.S. Carter, B. Kirkpatrick, I.I. Lederhendler (eds) The Integrative Neurobiology of Affiliation. *Annals of the New York Academy of Sciences* 807, 62–77.

230 **freezing** Porges has backed away somewhat from the use of this term. 'Freezing, although I have used the term, may be misleading. I now use freezing to describe immobilization with great muscle tone – and this reflects high sympathetic tone and prepares one to fight or flee, and I now use "death feigning" or shutdown, "vasovagal syncope" (fainting) to describe immobilization without muscle tone to describe the state supported by the old reptilian ANS.' Personal communication.

230 **fight-or-flight, and social engagement** Porges, S.W. (1995) Orienting in a defensive world: mammalian modifications of our evolutionary heritage. A Polyvagal Theory. *Psychophysiology* 32, 301–318.

231 **drowning themselves** Richter, C. (1957) On the phenomenon of sudden death in animals and man. *Psychosomatic Medicine* 19, 191–198.

231 **curse just placed on them will prove effective** Voodoo death is discussed in Bruce McEwen (2002) *The End of Stress as We Know It.* Washington: Joseph Henry Press; and Robert Sapolsky (2004) *Why Zebras Don't Get Ulcers* 3rd ed. New York: Henry Holt.

232 **inefficient cardiac output and high blood pressure** Blascovich, J., Tomaka, J. (1996) The Biopsychosocial Model of Arousal Regulation. *Advances in Experimental Social Psychology* 28, 1–51. The challenge-threat model and especially the claims that each attitude displays a distinct cardiovascular signature has been critically reviewed in Wright, R.A., Kirby, L.D. (2003) Cardiovascular correlates of challenge and threat appraisals: a critical examination of the biopsychosocial analysis. *Personality and Social Psychology Review* 7, 216–233.

232 **children with low vagal tone display more behavioural problems later in life** Porges, S.W., Doussard-Roosevelt, J.A., Portales, A.L., Greenspan, S.I. (1996) Infant regulation of the vagal 'brake' predicts child behavior problems: a psychobiological model of social behavior. *Developmental Psychobiology* 29, 697–712.

233 **correlated with a number of markers of health** See for example Sloan, R.P. et al. (2007) RR interval variability is inversely related to inflammatory markers: the CARDIA study. *Molecular Medicine* 13, 178–184.

233 **high ratio of anabolic to catabolic hormones** Huovinen, J. et al. (2009) Relationship between heart rate variability and the serum testosterone-to-cortisol ratio during military service. *European Journal of Sport Science* 9, 277–284.

234 **immune to the effects of stress and the stress hormones** Charney, D. (2004) Psychobiological mechanisms of resilience and vulnerability:

implications for successful adaptation to extreme stress. *American Journal of Psychiatry* 161, 195–216.

234 **more muted stress response to threats** Meaney, M., Aitken, D., van Berkel, C., Bhatnagar, S., Sapolsky, R. (1988) Effect of neonatal handling on age-related impairments associated with the hippocampus. *Science* 239, 766–768.

234 **18 per cent longer than that of non-stressed rats** Frolkis, V. (1981) *Aging and Life-Prolonging Processes*. Vienna: Springer-Verlag.

234 **ill-prepared to deal with the slings and arrows of normal life events** Meaney, M.J., Aitken, D.H., Viau, V., Sharma, S., Sarrieau, A. (1989) Neonatal handling alters adrenocortical negative feedback sensitivity and hippocampal type II glucocorticoid receptor binding in the rat. *Neuroendocrinology* 50, 597–604. Liu, D., Diorio, J., Tannenbaum, B., Caldji, C., Francis, D., Freedman, A., Sharma, S., Pearson, D., Plotsky, P.M., Meaney, M.J. (1997) Maternal care, hippocampal glucocorticoid receptors, and hypothalamic-pituitary-adrenal responses to stress. *Science* 277, 1659–1662.

235 **boot camp for the brain** Erickson, K. et al. (2011) Exercise training increases size of hippocampus and improves memory. *Proceedings of the National Academy of Sciences* 108, 3017–3022. Dienstbier, R., LaGuardia, R., Barnes, M., Tharp, G., Schmidt, R. (1987) Catecholamine Training Effects from Exercise Programs: A Bridge to Exercise-Temperament Relationships. *Motivation and Emotion* 11, 297–318. Foster, P., Rosenblatt, K., Kuljiš, R. (2011) Exercise-induced cognitive plasticity, implications for mild cognitive impairment and Alzheimer's disease. *Frontiers in Neurology* 2, 28, 1–15.

235 **sports science could help ... the person receiving it** Some standard references are the following: Jack H. Wilmore, David L. Costill (2004) *Physiology of Sport and Exercise* 3rd ed. Human Kinetics Publishers. Per-Olof Astrand, Kaare Rodahl, Hans A. Dahl, Sigmund B. Stromme (2003) *Textbook of Work Physiology* 4th ed. Human Kinetics Publishers. Frank W. Dick (2007) *Sports Training Principles* 5th ed. A. & C. Black Publishers Ltd.

235 **not as prone to learned helplessness** Dienstbier, R., Pytlik Zillig, L.M. (2005) Toughness. In C.R. Snyder & S. Lopez (eds) *Handbook of Positive Psychology*. New York: Oxford University Press. pp.512–527.

235 **enviable pattern of stress and recovery** Castellani, J., Degroot, D. (2005) Human endocrine responses to exercise-cold stress. In Kraemer, W., Rogol, A. (eds) *The Endocrine System in Sports and Exercise*. Oxford: Blackwell.

235 **sauna followed by a cold plunge** Hannuksela, M.L., Ellahham, S. (2001) Benefits and risks of sauna bathing. *American Journal of Medicine* 110, 118–126. Ohori, T. et al. (2012) Effect of repeated sauna treatment on

exercise tolerance and endothelial function in patients with chronic heart failure. *American Journal of Cardiology* 109, 100–104. A similar toughening practice of heating the body through exercise, followed by cold-water immersion, was recommended by the nineteenth-century German Sebastian Kneipp.

235 **later systems of emotional arousal** Stanley-Jones, D. (1966) The thermostatic theory of emotion: a study in kybernetics. *Progress in Biocybernetics* 3, 1–20.

235 **cold tolerance ... emotional stability** Dienstbier, R., LaGuardia, R., Wilcox, N. (1987) The Relationship of Temperament to Tolerance of Cold and Heat: Beyond 'Cold Hands-Warm Heart'. *Motivation and Emotion* 11, 269–295.

236 **weakened and inefficient by disuse** Walter Cannon (1932) *The Wisdom of the Body*. New York: Norton. pp.198–199.

236 **one cause of the current obesity epidemic** Keith, S. et al. (2006) Putative contributors to the secular increase in obesity: exploring the roads less travelled. *International Journal of Obesity* 30, 1585–1594.

237 **dropped below its metabolic cost** Boksem, M., Tops, M. (2008) Mental fatigue: costs and benefits. *Brain Research Reviews* 59, 125–139.

237 **no control over the allocation of their attention** Siegrist, J. (1996) Adverse health effects of high-effort/low-reward conditions. *Journal of Occupational Health Psychology* 1, 27–41. Van Der Hulst, M., Geurts, S. (2001) Associations between overtime and psychological health in high and low reward jobs. *Work Stress* 15, 227–240. Bosma, H., Marmot, M.G., Hemingway, H., Nicholson, A.C., Brunner, E., Stansfeld, S.A. (1997) Low job control and risk of coronary heart disease in Whitehall II (prospective cohort) study. *British Medical Journal* 314, 558–565.

238 **incredibly, outstanding personal achievement** Holmes, T.H., Rahe, R.H. (1967) The Social Readjustment Rating Scale. *Journal of Psychosomatic Research* 11, 213–218.

238 **later take a toll on their health** For a nice discussion of the effect of emotion, both positive and negative, on heart disease see Daniel Goleman (1995) Mind and Medicine. In *Emotional Intelligence*. London: Bloomsbury. Ch. 11.

240 **profound effect on vagal influences to the heart** Porges, Personal communication.

240 **rational choices in a financial decision-making task** Kirk, U., Downar, J., Montague, R. (2011) Interoception drives increased rational decision-making in meditators playing the ultimatum game. *Frontiers in Neuroscience* 5, 49.

242 **stressful life events and increased mortality** Rosengren, A., Orth-Gomér, K., Wedel, H., Wilhelmsen, L. (1993) Stressful life events, social

support, and mortality in men born in 1933. *British Medical Journal* 307, 1102–1105.

242 **the incidence of stress-related illness dropped noticeably** Robert A. Karasek, Tores Theorell (1992) *Healthy Work: Stress, Productivity and the Reconstruction of Working Life*. New York: Basic Books.

244 **Tortoise and Hare** John Coates, A Tale of Two Traders. *Financial Times* 4 May 2009.

CHAPTER 9: FROM MOLECULE TO MARKET

248 **'Black Swan' events** Nassim Nicholas Taleb (2007) *The Black Swan: The Impact of the Highly Improbable*. New York: Random House.

249 **doubts about ... rational choice** See for example John Maynard Keynes (1938) My Early Beliefs. In *Essays in Biography: The Collected Writings of John Maynard Keynes, Vol. X*. London: Macmillan. Robert Skidelsky (1995) *John Maynard Keynes: The Economist as Saviour, 1920–1937*. London: Penguin.

249 **Here we are off any map drawn by rational-choice theory** However, behavioural economists have suggested ingenious management and policy innovations. See Richard H. Thaler, Cass R. Sunstein (2008) *Nudge: Improving Decisions About Health, Wealth, and Happiness*. Yale University Press; Hersh Shefrin (2008) *Ending the Management Illusion: How to Drive Business Results Using the Principles of Behavioral Finance*. New York: McGraw-Hill.

250 **mental illness while holding office?** Owen, D., Davidson, J. (2009) Hubris syndrome: An acquired personality disorder? A study of US Presidents and UK Prime Ministers over the last 100 years. *Brain* 132, 1407–1410. This theme is developed in a fascinating book, David Owen (2008) *In Sickness and in Power: Illness in Heads of Government During the Last 100 Years*. London: Methuen. Owen has set up a research trust, called The Daedalus Trust, to research disorders stemming from the exercise of power in both politics and business.

251 **market stability needs biological diversity** John Coates, Traders Should Track Their Hormones. *Financial Times* 14 April 2008.

252 **'tend-and-befriend' reaction, an urge to affiliation** Shelley E. Taylor et al. (2000) Biobehavioral Responses to Stress in Females: Tend-and-Befriend, not Fight-or-Flight. *Psychological Review* 107, 411–429.

252 **more stressed by social problems, with family and relationships** Stroud, L., Salovey, P., Epel, E. (2002) Sex differences in stress responses: social rejection versus achievement stress. *Biological Psychiatry* 319, 318–327. Katie T. Kivlighana, Douglas A. Granger, Alan Booth (2005) Gender differences in testosterone and cortisol response to competition. *Psychoneuroendocrinology* 30, 58–71. R.E. Bowman (2005)

Stress-Induced Changes in Spatial Memory are Sexually Differentiated and Vary Across the Lifespan. *Journal of Neuroendocrinology* 17, 526–535.

252 **women may be less hormonally reactive than men** Coates, J., Gurnell, M., Sarnyai, Z. (2010) From molecule to market: steroid hormones and financial risk-taking. *Philosophical Transactions of the Royal Society* B 365, 331–343.

253 **women are more risk-averse than men** Eckel, C., Grossman, P.J. (2008) Men, women and risk aversion: Experimental evidence. In *Handbook of Experimental Economic Results, Vol.1.* New York: Elsevier. Powell, M., Ansic, D. (1998) Gender differences in risk behavior in financial decision-making: an experimental analysis. *Journal of Economic Psychology* 18, 605–628. Schubert, R., Brown, M., Gysler, M., Brachinger, H.W. (1999) Financial decision-making: Are women really more risk-averse? *American Economic Review* 89, 381–385. Rachel Croson and Uri Gneezy (2009) Gender Differences in Preferences. *Journal of Economic Literature* 47, 1–27.

253 **single women outperformed single men by 1.44 per cent** Brad Barber, Terrance Odean (2001) Boys will be Boys: Gender, Overconfidence, and Common Stock Investment. *Quarterly Journal of Economics* 261–295.

255 **how to stop men from seceding from it** Andrew Sullivan, The He Hormone. *New York Times Magazine* 2 April 2000.

259 **If the eye were an animal, its soul would be seeing** Aristotle, *De Anima* 412b, 18–19.

260 **unified policy science, from molecule to market** McEwen, B. (2001) From molecule to mind: stress, individual differences, and the social environment. In *Unity of Knowledge: The Convergence of Natural and Human Science* eds A. Damasio et al. *Annals of the New York Academy of Sciences* 935, 42–49. See as well Matt Ridley (2004) *Nature via Nurture: Genes, Experience and What Makes us Human.* London: Harper Collins.

260 **economics beginning to merge … with epidemiology** There already exists a branch of study known as economic epidemiology. See for example Philipson, T. (2000) Economic epidemiology and infectious disease. In Cuyler, A., Newhouse, J. (eds) *Handbook of Health Economics.* Amsterdam: North Holland. pp.1761–1799. An excellent piece of economic epidemiology can be found in Gerdtham, U., Johannesson, M. (2005) Business cycles and mortality: results from Swedish microdata. *Social Science and Medicine* 60, 205–218.

260 **two cultures** Snow, C.P. (1959) *The Two Cultures.* London: Cambridge University Press.

FURTHER READING

INTRODUCTION

Joe LeDoux (1996) *The Emotional Brain: The Mysterious Underpinnings of Emotional Life*. New York: Touchstone. A classic work, written by a neuroscientist who has participated in much of the research I refer to in this book.

Coates, J., Gurnell, M., Sarnyai, Z. (2010) From molecule to market: steroid hormones and financial risk-taking. *Philosophical Transactions of the Royal Society* B 365, 331–343. A review article written by myself and two colleagues, Dr Mark Gurnell, an endocrinologist, and Zoltan Sarnyai, a pharmacologist. First half is a survey of research at the cellular level, second half the behavioural level.

BODY AND MIND

Damasio, Antonio (1992) *Descartes' Error*. New York: Avon Books. Damasio's book on how he and Antoine Bechara discovered the essential role of bodily signals in rational thought.

Sandra Blakeslee, Matthew Blakeslee (2007) *The Body Has a Mind of its Own: How Body Maps in Your Brain Help You Do (Almost) Anything Better*. New York: Random House.

Craig A.D. (2002) How do you feel? Interoception: the sense of the physiological condition of the body. *Nature Reviews Neuroscience* 3, 655–666. An important review article. A more accessible version of Craig's work can be found in a recorded lecture: The neuroanatomical basis for human awareness of feelings from the body. Found at http://vimeo.com/8170544

BEHAVIOURAL AND NEURO-ECONOMICS

Hersh Shefrin (2007) *Beyond Greed and Fear: Understanding Behavioral Finance and the Psychology of Investing*. Oxford University Press. Excellent overview of the field.

Richard H. Thaler, Cass R. Sunstein (2008) *Nudge: Improving Decisions About Health, Wealth, and Happiness*. Yale University Press. Influential book on how behavioural economics can inform economic policy.

Daniel Kahneman (2011) *Thinking, Fast and Slow*. New York: Farrar, Straus & Giroux. A summary of Kahneman's work.

Paul Glimcher, Ernst Fehr, Antonio Rangel, Colin Camerer, Russell Poldrack (2009) *Neuroeconomics: Decision-Making and the Brain*. London: Academic Press.

Camelia Kuhnen, Brian Knutson (2005) The Neural Basis of Financial Decision-Making. *Neuron* 47, 763–770.

HOMEOSTASIS

Walter Cannon (1932) *The Wisdom of the Body*. New York: Norton. Dated but well-written masterpiece by the scientist who discovered homeostasis and the fight-or-flight response.

Thomas Amini, Fari Lannon, Richard Lewis (2001) *A General Theory of Love*. New York: Vintage. Lovely little book on homeostasis and love.

VISUAL SYSTEM AND SPEED OF REACTIONS

Tor Norretranders (1991) *The User Illusion*. New York: Penguin.

Tom Stafford and Matt Webb (2005) *Mind Hacks: Tips and Tools for Using Your Brain*. Sebastopol, CA: O'Reilly Media. Wonderful, fun book that explains many of the tricks our brain uses to simplify our understanding of the world, and speed up our reactions. Each chapter has an easy experiment you can perform to observe your brain at work.

Stephen Pinker (1999) *How the Mind Works*. New York: Norton.

Ken Dryden (2003) *The Game*. New York: Wiley. On conscious and unconscious mind in sport, written by a great athlete.

GUT FEELINGS

Antoine Bechara, Antonio R. Damasio (2005) The somatic marker hypothesis: A neural theory of economic decision. *Games and Economic Behavior* 52, 336–372.

Timothy Wilson (2002) *Strangers to Ourselves: Discovering the Adaptive Unconscious*. Boston: Harvard University Press.

Malcolm Gladwell (2005) *Blink: The Power of Thinking Without Thinking*. New York: Little, Brown.

VAGUS NERVE AND ENTERIC NERVOUS SYSTEM

Stephen Porges (2011) *The Polyvagal Theory: Neurophysiological Foundations of Emotions, Attachment, Communication, and Self-Regulation*. New York: Norton. A collection of articles by the scientist who developed the theory of the vagal brake.
Michael Gershon (1998) *The Second Brain*. New York: HarperCollins. A highly readable book on the enteric nervous system.

SEARCH

Gregory Berns (2006) *Satisfaction: Sensation Seeking, Novelty, and the Science of Finding True Fulfillment*. New York: Henry Holt. On dopamine, written in a conversational style.
Donald Pfaff (2005) *Brain Arousal and Information Theory: Neural and Genetic Mechanisms*. Boston: Harvard University Press. A cutting-edge treatise, written for scientists.

TESTOSTERONE AND IRRATIONAL EXUBERANCE

James M. Dabbs (2000) *Heroes, Rogues and Lovers: Testosterone and Behavior*. New York: McGraw-Hill. An overview of research on testosterone and behaviour.
John Maynard Keynes (1936) 'The State of Long-Term Expectation' in *The General Theory of Employment, Interest and Money*, Chapter 12. London: Macmillan. This one chapter in Keynes's great work is the single best description of exuberance, a must-read.
Robert Shiller (2005) *Irrational Exuberance* 2nd ed. Princeton University Press.
George A. Akerlof and Robert J. Shiller (2009) *Animal Spirits: How Human Psychology Drives the Economy, and Why It Matters for Global Capitalism*. Princeton University Press.
Michael Lewis (1990) *Liar's Poker: Rising Through the Wreckage on Wall Street*. New York: Penguin. Still the best description of trading-floor bravado.
David Owen (2008) *In Sickness and in Power: Illness in Heads of Government During the Last 100 Years*. London: Methuen. An original account of hubris and clinical conditions in political leaders, written by a senior British politician who also happens to be a trained neurologist.

STRESS

Bruce McEwen (2002) *The End of Stress as We Know It*. Washington: Joseph Henry Press. A review of stress research by one of the greats in the field.
Robert Sapolsky (2004) *Why Zebras Don't Get Ulcers* 3rd ed. New York: Henry Holt. A book that covers all aspects of stress, written by one of the scientists who, together with McEwen, has done so much of the research on stress and the brain.
Robert A. Karasek, Tores Theorell (1992) *Healthy Work: Stress, Productivity and the Reconstruction of Working Life*. New York: Basic Books.

TOUGHNESS

Dienstbier, R.A. (1989) Arousal and physiological toughness: Implications for mental and physical health. *Psychological Review* 96, 84–100. The classic article on toughness.

SPORTS SCIENCE

Jack H. Wilmore, David L. Costill (2004) *Physiology of Sport and Exercise* 3rd ed. Human Kinetics Publishers.
Per-Olof Astrand, Kaare Rodahl, Hans A. Dahl, Sigmund B. Stromme (2003) *Textbook of Work Physiology* 4th ed. Human Kinetics Publishers.
Frank W. Dick (2007) *Sports Training Principles* 5th ed. A. & C. Black Publishers Ltd.

MISCELLANEOUS

Brian Brett (2004) *Uproar's Your Only Music*. Exile Editions. Lovely, uncategorisable book in which the author tells of growing up with Kallmann syndrome, a rare genetic disorder which left him without testosterone.
Matt Ridley (2004) *Nature via Nurture: Genes, Experience and What Makes us Human*. London: HarperCollins. An overview of the nature/nurture debate.

INDEX

303

effect of steroid hormones on 23–8
evolutionary model 35–7, 48
feedback 28–9, 48, 92–7, 100–7
and homeostasis 21–3, 44–8
and the hypothalamus 23, 46–7,
151, 198, 202, 208, 221, 223
and information processing 125–9
and interoception 47–8
and last in, first out rule 92
and listening to our body 111–13,
237–40
and memory 208–10
and neo-cortex 41
philosophic views 28–9
and physical movement 37–41
revving up 42–3
and selective attention 91–3
and stress response 207–17
and thermoregulation 45
unity of 33–4

bonus payments 5, 10, 12, 32, 57, 79,
87–8, 148–50, 164, 176, 179, 182,
184, 187, 206, 212, 244–6, 255
Borg, Björn 170
boxing 61–2, 72, 119, 170
brain stem 7, 44, 46, 71, 102, 124,
135–6, 191, 193, 208–9, 221, 239
British United Provident Association
Ltd 216
Brown-Séquard, Charles Edward
152, 154
Buddhist meditation 240
Buffett, Warren 251
bull market
and market bubbles 13–14, 16–18
and testosterone 20–1, 24–5, 27
trading floor example 54, 75, 106,
131, 149–52, 164–5, 180
unsustainable 54
and the winner effect 166, 177,
179–80
Bush, George W. 165

Butenandt, Adolf 153–4

Camerer, Colin 83, 270, 300
Cannon, Walter 95–6, 236
Case-Shiller index 188
catabolic hormones 226
cerebellum 45, 122–3, 209
and body–brain 41–2
Champagne, J. 277
Chicago Board Options Exchange
(VIX) 129
Chicago Board of Trade 76–8
Chicago Mercantile Exchange 150
Chiel, H. 265
Chrousos, G. 292
Churchill, Winston 3
Churchland, Patricia 269
Clark, Andy 38, 265
cocaine 20. 136, 138, 142
Cocktail Party Effect 125
Condon, William 97–8
consciousness 59, 62–8, 73–4
cortisol
and anxiety 229
and body–brain 9, 22
and stress response 26–7, 106, 144,
192, 198–205, 207–15, 217, 221–2,
224–30, 232, 235, 240–1, 251
Coward, Noël, *Design for Living* 153
Craig, Bud 47–8
Credit Crisis (2007–08) 80, 88–9,
129, 206
CRH (corticotropin-releasing
hormone) 211–12
cricket 60, 62–4, 70, 72
Critchley, H.D. 266, 275
Cushing's syndrome 210, 216

Dagher, Alain 277
Damasio, Antonio 48, 90, 91, 108–9,
271, 274, 298, 299, 300
Davis, Greg 261, 281
Dayan, Peter 276

stress response 43, 191–5
and bear market 198–207
controlling 221–3
and novelty, uncertainty,
uncontrollability 202–4, 238–42
workplace suggestions 240–7
stress-related disease 213–17
sub-prime mortgages 186–8, 190
Sullivan, Andrew 26, 255
superior colliculus 69
Sykes, Bryan 155, 279

Taleb, Nassim 248, 297
Taylor, Shelley 252
tend-and-befriend 252, 297
tennis 48, 58, 60, 66, 107, 119, 145,
168–70
testosterone 9, 24–7, 115, 151–8
2D:4D ratio 174–5
age differences 251, 255
anabolic/masculinising effects of
157–8
description of 154–7
discovery of 152–3
effect on sex drive 156, 215
fluctuations in 155–6, 171–2
and foetal exposure 156–7
rising/elevated 177–80
role of 255–6
used to treat andropause 153
and winner effect 167–75
testosterone feedback loop 25, 172–5,
179
thalamus 192–3
Thaler, Richard 30, 264, 270, 297, 300
Theorell, Tores (with Robert
Karasek), Healthy Work 216, 242
thermoregulation 45, 235–6
thriving 226–7
tickling 123, 276
Tiger Capital 13
toughness
and amines 227–30

and anabolic hormones 226–7
and catabolic hormones 226
challenge and threat 233–4
exposure to coldness 235–6
mental 223–34
physical 223, 234–6
training of 224–5, 234–7
and vagus nerve 230–2
trading floor example
arb (arbitrage) desk 159–65
bear market stress response
198–207
boredom/dullness 5
and bull markets 54, 75, 106, 131,
149–52, 164–5, 180
changes in hormone levels 9
confusion then calm 147–8
cycle of euphoria-excessive risk-
taking-crash 10
and dopamine–testosterone mix
151–2
early-warning system 6–7, 121,
125–6
excitement/volatility 7–8
fast reactions 75–81
and feedback loops 101–7
fight-or-flight response 191–5,
198–9
flexibility of traders 79–81
high-frequency trading 77, 80
and learned helplessness 213–14
listening to faint voices 111
and market exuberance 149–52,
176–7
market information/trader arousal
128–35
and the orienting response 145–6
and pattern recognition/predicting
the market 78, 85–9, 110, 113–14,
127, 133, 211
physical nature of 76–7
and physiological monitoring 115
recovery period 119–21